Wednesday Is Indigo Blue

Also by Richard E. Cytowic, M.D.

Synesthesia: A Union of the Senses (2nd edition)
Nerve Block for Common Pain
The Neurological Side of Neuropsychology
The Man Who Tasted Shapes
Prepare to Say No: Mastering the Art of Personal Choice
The Magician's Accomplice: My Father and I in the Age of Anxiety

Also by David M. Eagleman, Ph.D.

Dethronement: The Unconscious Brain Behind the I
Plasticity: How the Brain Reconfigures Itself on the Fly
Sum: Tales from the Afterlives

WEDNESDAY IS INDIGO BLUE

Discovering the Brain of Synesthesia

Richard E. Cytowic, M.D., and David M. Eagleman, Ph.D.

A Bradford Book

The MIT Press
Cambridge, Massachusetts
London, England

First MIT Press paperback edition, 2011
© 2009 Massachusetts Institute of Technology
Afterword Copyright © 2009 Dmitri Nabokov

MIT Press books may be purchased at special quantity discounts for business or sales promotional use. For information, please e-mail special_sales@mitpress.mit.edu.

This book was set in Stone Sans and Stone Serif by SNP Best-set Typesetter Ltd., Hong Kong and was printed and bound in the United States of America.

Library of Congress Cataloging-in-Publication Data
Cytowic, Richard E.
Wednesday is indigo blue : discovering the brain of synesthesia / Richard E. Cytowic and David M. Eagleman.
    p. cm.
Includes bibliographical references and index.
ISBN 978-0-262-01279-9 (hardcover : alk. paper)—978-0-262-51670-9 (pb.) 1. Synesthesia. 2. Emotions and cognition. I. Eagleman, David. II. Title.
RC394.S93C964   2009
612.8'233—dc22
                                                                    2008029814

10 9 8

Contents

Introduction vii

1 What Color Is Tuesday? 1

2 A Kaleidoscopic World 23

3 Don't It Make My Brown I's Blue? 63

4 See with Your Ears 87

5 November Hangs above Me to the Left 109

6 A Matter of Taste 127

7 Auras, Orgasms, and Nervous Peaches 151

8 Metaphor, Art, and Creativity 163

9 Inside a Synesthete's Brain 199

10 Questions Ahead 235

Afterword by Dmitri Nabokov 249
Notes 255
Bibliography 281
Index 301

Introduction: Your Neighbor's Reality

Erica Borden watches weather patterns on a giant screen each day. She is a twenty-seven-year-old meteorologist with an expressive face framed by long dark hair. The weather patterns are shaping up for a beautiful day, and at the moment Erica is sharing a package of chocolate-covered raisins with her colleague Aviva. She has no reason to suspect there is anything unusual about her brain.

But from the day of her conception, a miniscule genetic change lurked deep in her chromosomes. This tiny change, expressing itself in her billions of brain cells, makes her reality different from her friend's.

Aviva sits nearby and pops a raisin in her mouth, also not suspecting that she and her friend experience a different world. Once, in college, she asked her roommate, "How do I know I see red the same way you see red? What if what you call red is what I call green?" They concluded they cannot know, but that it didn't matter how they saw it on the inside so long as they both agreed to call an apple "red." To this day Aviva continues to be intrigued by the idea that two people may see red differently. She doesn't know that the possibility of dissimilar experience goes much deeper than color, and she doesn't know that Erica's reality is measurably different than hers.

And so Erica and Aviva blithely sit next to one another sharing their raisins. When Erica tastes a chocolate-covered raisin on her tongue, she feels a nubbled texture on her fingertips. When she hears the voice of the weather announcer, she can't help but sense a deep indigo color that ripples in the upper left corner of her visual field. When she thinks of today—Thursday—the concept seems to occupy a particular region of space near her right shoulder. Erica's brain is like the weather system on the coast: all the elements interact because there are no barriers to keep them from mixing. Erica's senses and concepts are open to each other, flowing and merging like weather streams.

On the other hand, things are neatly compartmentalized in Aviva's brain. A raisin is only a raisin. Voices are heard, not seen. Thursday makes her anticipate the weekend, but it has no location. Her brain is like the weather pattern in the rocky mountainous regions: isolation by a mountain makes weather in one spot independent of weather in the next range.

Erica and Aviva have no idea they perceive the world differently.

Most people's brains are compartmentalized like Aviva's.

Erica, on the other hand, enjoys an unusual condition called synesthesia.

Wednesday Is Indigo Blue

1   What Color Is Tuesday?

Most people have never heard of synesthesia.

Yet everyone knows the word "anesthesia," meaning "no sensation." Rhyming with it and sharing the same root (Greek *syn* = union + *aisthaesis* = sensation), "synesthesia" means "joined sensation," such that a voice or music, for example, is not only heard but also seen, tasted, or felt as a physical touch. Some individuals with synesthesia are shocked to discover as children that the rest of the world does not experience things as they do. Many other synesthetes reach adulthood completely unaware that their experience is in any way unusual.

"I thought everybody—!" they exclaim.

It is often by accident that synesthetes discover that their way of perceiving is not the norm. As a seven-year-old, for example, synesthetic artist Carol Steen once said to a schoolmate, "The letter 'A' is the most beautiful pink I've ever seen."

Thinking she must be crazy, Carol's classmate gave her a withering look. After that, Carol never mentioned her colors to anyone until she was twenty, when, as she and her family were sitting around the dinner table one evening, she told them that the number 5 was yellow—whereupon her father startled her by insisting, "No, it's yellow ochre!" and then refused to say anything more about it. Later, when she was in her thirties and teaching art at the University of Michigan, a psychology colleague revealed that this experience of colored letters had a name: synesthesia.

At the time, all Carol could uncover about synesthesia was a dictionary definition, and then it only mentioned colored music. She went twenty more years starved for information until she happened to hear one of the authors of this book, Richard Cytowic, explaining synesthesia on National

Public Radio.[1] Only then did she learn that her experience was not only genuine but also scientifically important. Forever grateful, she now says, "The knowledge he shared gave me my freedom after fifty years of isolation."

Another woman with what are called "number forms" reached college before realizing that most people do not sense numerals along a three-dimensional line that twists and zigzags in space. She complained to her math professor of having difficulty with the equations "because the digits keep going up to their places." Sensing something significant in her offbeat remark, the professor handed her a length of stiff wire and asked her to indicate where along it the figures were located. He then watched as, without the least hesitation or surprise, "she took the wire and bent it here and there in three dimensions until it looked like a tortured thing. A number of times, she returned to previously made bends, correcting the angles precisely."[2]

"Is anything odd about it?" the student asked in all seriousness. "Everybody sees numbers like that, don't they?"

She was stunned to learn that they did not.

Some synesthetes live a very long time before discovering their exceptional ways. Jean Milogav, a Swiss synesthete, reached her sixties before realizing that sensing the alphabet and numerals as colored was rather unusual:

I only became consciously aware of it after reading Vladimir Nabokov's autobiography *Speak, Memory* in which, to my great surprise, he describes exactly the same type of synesthesia as mine: i.e., "seeing" every letter of the alphabet and every number in a specific color. . . . My colors are not the same as his.

I too never talked about it, not out of shyness but because I always thought all people were like that. Only after reading Nabokov's description of his synesthesia did I realize this was rather unusual. . . . I enjoy it very much and would be hard put if these colors would suddenly vanish. I don't think they will; I am 61 now and had it all my life.[3]

Jean speaks German, Italian, French, and Spanish, and because her colors are determined by the physical appearance of each letter rather than by their sound, it is the spelling that determines the color of a word no matter which language she speaks.

As noted above, accident is how synesthetes often first learn about their astonishing gift, and accident also lay behind Richard's rediscovery of

Figure 1.1
In Michael Watson, taste and smell triggered tactile sensations of contour, texture, weight, and temperature.

synesthesia in February 1980. His dinner host delayed their seating with the apology that there weren't "enough points on the chicken." Michael Watson, literally *The Man Who Tasted Shapes* in the book of that title that Richard would later write, felt taste and smell as a physical touch on his face and in his hands.

"With an intense flavor," he explained, "the feeling sweeps down my arm into my hand and I feel shape, weight, texture, and temperature as if I'm actually grasping something" (see figure 1.1). He had wanted the chicken to be a prickly, pointed sensation, "like laying my hand on a bed of nails." But it came out all round.

"I can't serve this," he insisted, embarrassed at having roasted the wrong shape. Michael enjoyed cooking, but rather than flavor being his gourmet guide he liked to conjure up dishes that evoked particular shapes, textures, and other tactile sensations.

"They're Just Looking for Attention"

At that time, no one in Richard's academic circle had ever heard of synesthesia, and science had lost interest in it for many decades because it couldn't explain it or even verify someone's subjective experience. Richard only knew the term because he had read a translated book called *The Mind of a Mnemonist* by the famous Soviet neuropsychologist A. R. Luria, describing a memory expert named Sheresevsky, who remembered limitless amounts thanks to a fivefold synesthesia that spanned his senses. For example, a bell triggered seven simultaneous perceptions for Sheresevsky:

I *heard* the bell ringing . . . a *small, round* object *rolled* before my eyes . . . my fingers sensed something *rough* like a rope . . . I experienced a *taste* of salt water . . . and something *white*.[4]

Richard's neurology colleagues joked that his subject, Michael, had to be either crazy or on drugs, insisting that synesthesia could not be a real perceptual phenomenon because it contradicted standard notions of separate sensory channels in the brain. They warned him not to pursue it scientifically because it was "too weird, too New Age," and "would ruin" his career. In other words, they had the typical reaction of orthodoxy to something it cannot explain: deny it and sweep it under the carpet.

Thanks to Richard's pioneering work, however, the paradigm has shifted, and today young researchers around the world are writing articles, Ph.D. theses, and scholarly books and papers about synesthesia and are having to think about the brain in ways they have never had to think about it before. You will meet many of them in these pages.

For a long time, though, it was typical to dismiss synesthesia with glib or even hostile assertions. "They're just imagining it," skeptics claimed, writing synesthetes off as needy exhibitionists with overactive imaginations who simply want to call attention to themselves. One feature that skeptics grabbed onto in support of their contention that synesthetes were just "making it up" is the fact that synesthetic associations are idiosyncratic. That is, no two synesthetes, not even identical synesthetic twins, experience the same colors if given the same stimulus. And the assertion that synesthetes were "on drugs" was not entirely arbitrary, because LSD and mescaline sometimes *do* cause synesthesia, both during the high and afterwards. However, drug-induced synesthesia is a different experience from the naturally occurring state. (Yet the fact that drugs can induce a similar state only makes the naturally occurring kind more intriguing.) When other excuses failed, critics ultimately dismissed synesthetes as "crazy artists."

Many in the scientific establishment rolled their eyes at the notion of synesthesia and tried to explain away a very common type—namely, sensing letters and numbers as colored even though they are printed in black ink—by arguing that synesthetes were just "remembering" childhood associations from coloring books or refrigerator magnets, and that is why the letter A was red or D was green, for example. But synesthesia runs strongly in families, a fact Sir Francis Galton noted more than a century ago in Britain. It would be an implausible coincidence if subsequent family generations all happened to "remember" color associations from magnets handed down. Moreover, almost all of us played with coloring books and refrigerator magnets as youngsters, but for the majority a particular color "memory" does not become irreversibly bound to digits or letters. Finally, each synesthete's palette of color pairings is quite idiosyncratic—a fact true even between family members. Therefore, a privileged memory is likely not the explanation for synesthesia.

What Galton astutely noted, actually, is that otherwise normal people saw colors whenever they *looked* at numbers or letters. That is, it was the visual appearance of the written grapheme that triggered an experience of color. By contrast, we note that the sounds of language, called *phonemes*, tend to trigger synesthetic tastes. For James Wannerton, the word "village" tastes like sausage, as do words with the same /idg/ sound such as "college" and "message." The name "Derek" tastes like earwax, and "safety" like lightly buttered toast. It further turns out that phonemes triggering tastes tend to be present in the word for that taste. Thus, "April" tastes of apricots, "Barbara" like rhubarb, and "Cincinnati" like cinnamon rolls. Childhood food names apparently act as templates for other words triggering synesthetic tastes, a finding whose importance we elaborate in chapter 6.

A remarkable observation is that once a synesthetic link is established, it seems to remain stable for life. That is, for a given individual, the letter A is always cobalt blue or "Derek" always tastes like earwax. Once established, the association becomes locked in. Most fascinating, synesthesia is genetically inherited and thus a function of nature but also requires interaction with nurture in the form of early life exposure to culturally learned artifacts such as letters, numbers, and food categories. In fact, permanent links get established between aspects of sensation and a whole range of concepts as shown by the many varieties of synesthesia—because basically the gene confers increased cross talk between different brain areas. Whether the cross talk results from extra wiring or more activity in the normal wiring is the subject of chapter 9.

In the case of colored letters and numbers, the brain area crucial for recognizing letter and numeral graphemes (the green area in figure 1.2) is positioned in the left hemisphere just next to the color perception area (the red area in figure 1.2) called V4. Because of increased cross talk, the appearance of a letter triggers activation in the V4 color area. It is interesting that despite having normal color vision, synesthetes commonly say they experience "weird" or "ugly" colors they would not deliberately choose. Even Steven S., a partially color-blind synesthete, speaks of seeing "Martian colors."[5] Some of his photoreceptors are abnormal, restricting the range of colors he can see, but the color areas in his brain appear to be driven by alternate nonoptical inputs. Cases like this rule out the argument that synesthesia is nothing more than childhood memories, because how could someone remember colors he or she has never seen or is incapable of seeing?

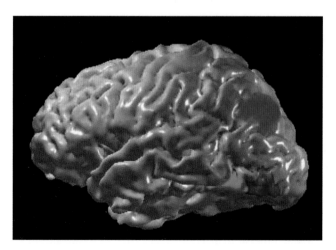

Figure 1.2
The brain area for recognizing letter and numeral graphemes (green) is positioned in the left hemisphere just next to the color perception area (red) called V4.

A recurrent criticism of skeptics is that synesthetes are simply speaking metaphorically, the way someone might speak of a "loud tie."[6] But think a minute—isn't a tie visual instead of auditory? And why do people use a taste adjective to describe a person, as in "she's so sweet?" What is going on with terms like "sharp cheese" and "cool jazz"? There is a circular logic in saying synesthetes are just being metaphoric because we do not yet understand how metaphor is represented in the brain. Rather, the argument should perhaps go the other way around: perhaps common metaphors stem from synesthesia. Our hope is that understanding the concrete sensory phenomenon of synesthesia will give us a handle on the neurological basis of metaphor and even artistic creativity. We explore this relationship in chapter 8.

More Common than Originally Suspected

Given that most people have never heard of synesthesia, one of the first questions tends to be "How common is it?"

In 1880, Sir Francis Galton[7] assumed the prevalence of synesthesia to be 1 in 20 based on his observations of individuals with "visualized numerals." His contemporaries came up with similar estimates ranging from 1 in 4 to 1 in 10.[8] By contrast, Richard studied numerous varieties of perceptual synesthesia and, in 1989, estimated synesthesia in general to be quite rare at 1 in 25,000. He extrapolated from known cases in North America. An informal follow-up survey of two million online individuals in 1994 agreed with that figure.[9]

A drawback of the survey method is that it is far from the ideal of a random sample of the population. All survey samples are prone to bias. For example, while two million individuals is an impressive pool of candidates, it includes only people who own computers and subscribe to a particular Internet provider. This methodology further skews results by selecting only those who read the survey and by depending on individuals to come forward and identify themselves as synesthetes rather than having a researcher systematically examine each one to ascertain who really has the trait.

As synesthesia gradually became a respectable topic for scientific study again, other researchers aimed to pin down its prevalence more rigorously. In 1993, Simon Baron-Cohen and colleagues in London surveyed two nonrandom populations and came up with estimates of 1 in 2,000 and 1 in 2,500.[10] They derived their estimates from the number of respondents to newspaper advertisements in campus and town papers, basing their calculations on the papers' circulation. Although also depending on self-referrals, the team did personally evaluate positive responses to determine whether those individuals truly were synesthetic. Because the entire participant pool was not questioned, however, no conclusions could be drawn about those who did not respond. Thus, the estimate of 1 in 2,000 was conservative.

As synesthesia became more widely known among neuroscientists, it also drew increasing attention in the popular press. Accordingly, synesthetes began to read about themselves in newspapers and magazines or hear about themselves on radio programs, and they eagerly contacted the researchers being interviewed. More estimates of the prevalence of synesthesia became available based on populations who came forward as self-identified synesthetes available for study. In Germany, Hinderk Emrich and colleagues estimated synesthesia to occur in between 1 in 300 and 1 in 700 individuals,[11] whereas in the United States, Vilayanur Ramachandran and Edward Hubbard put the prevalence at 1 in 200 based on classroom surveys.[12] Since these estimates were again based on self-report, they shared the same shortcomings as earlier ones.

Finally, in 2005, Julia Simner and colleagues in Edinburgh tested two populations, one at a university and another visiting a large science museum. Her team overcame earlier shortcomings by individually assessing a large random sample of people and verifying their self-reports with objective tests. The university study tested for all known types of synesthesia, whereas the museum study asked only about colored letter synesthesia. Their calculation is therefore the most sound to date because of its sampling methods, the use of objective tests, and the agreement of both estimates using two different populations and testing methods.

The results confirmed that synesthesia is far more common than originally assumed: 1 in 23 for any type of synesthesia, and 1 in 90 for grapheme → color synesthesia. It was also determined that the most common type of synesthesia was experiencing color for days of the week, followed by grapheme → color synesthesia. Previously, the latter was widely believed to be the most common manifestation.

We relate this history to illustrate how scientific "facts" shift based on new information. Typically, answering one question raises others. Sometimes a shift occurs as a result of better methodology or accuracy. Other times it results from asking different questions. Often, we are left with loose ends or even data that conflict.

A good example of this is the striking observation in the earlier studies of a strong-overrepresentation-of-females bias among synesthetes: in the past, three to six times as many women reported having synesthesia as men. This finding seemed to have strong theoretical implications, suggesting a mode of genetic inheritance that is linked to the X chromosome. That is, mothers (who possess two X chromosomes) can pass an X gene on to either sons (XY) or daughters (XX), while fathers (having XY chromosomes) can transmit the gene only to daughters. People therefore began to suspect that the genetic basis of synesthesia would be associated with the X chromosome.

However, that hypothesis was still not enough to explain the massive outnumbering of women to men. The problematic observation of a ratio as high as 6 to 1 was addressed by appealing to a genetic event known as "male lethality." The idea was that genes associated with synesthesia were deadly to half of male embryos, who would accordingly miscarry, leaving an excess of female synesthetes among live births. Thus, the excess number of female synesthetes seemed accounted for.

It was therefore a surprise when Julia Simner's randomly sampled population showed no female bias. Instead, she found roughly equal numbers of synesthetic men and women. The conclusion is that the strong female biases detected by earlier studies were all likely due to sex differences in self-disclosure. That is, women are far more likely than men to come forward and report interesting and unusual experiences such as synesthesia. As a result of Simner's study, we can drop the need to hypothesize maternal inheritance and male lethality.[13] One of us, David, has for years been studying synesthetic families and their DNA, and in chapter 9 we turn in detail to the genetic findings and what they teach us about the neural basis of synesthesia.

An Early Family Affair

Synesthesia is usually evident at an early age. Individuals invariably claim to have had it as far back as they can remember and cannot recall ever not having it. Indeed, a childhood assumption is that everyone is like them.[14]

When they realize this is not the case, the pendulum may swing the other way, making individuals then believe they are the only person in the world who perceives as they do. For example, Bruce Brydon, who has emotionally mediated synesthesia causing him to see colored auras around objects, felt isolated:

I have never communicated to anybody of seeing additional colored light. For one, I have failed to understand it myself, and to try to explain it to somebody else would leave me no better off. I was so happy to see that my experience is shared and acknowledged by others. I'm 35 years old and work in the construction industry. Fear of ridicule has held my secret.

Unfortunately for young synesthetes, disbelief and ridicule are real possibilities. Deni Simon has three types of synesthesia, hearing → color, grapheme → color, and emotion → color. She recalls:

My parents thought I was very strange. They thought I was making it up to get attention. Everyone was always jumping in with psychological explanations: I had an overactive imagination, I was spoiled and wanted attention, a whole slew of things. My mother was the only person that believed me, and I'm not sure she was truly convinced that what I experience is real.

Sometimes, untoward results occur. For example, a baby-sitter alarmed a four-year-old's parents by insisting, "Schuyler is psychotic!" He had told her that "colored straws" were floating in his apple juice (possibly indicating smell/taste causing an experience of color + shape). He further made endless crayon drawings of sounds such as a "helicopter sound" or the "cuckoo clock sound" (possibly indicating colored hearing). Fortunately, both parents were Ph.D. students, and neither gave in to the sitter's excitement. Rather, they scoured their university's library for alternative explanations and eventually rang up Richard to inquire whether their child might be synesthetic.

As Galton had noted, synesthesia tends to run in families.[15] When it does, otherwise odd remarks are thankfully recognized for what they are. For example, when Vladimir Nabokov, as a toddler, complained to his mother that the colors on his alphabet blocks were "all wrong," she understood him to mean that the colors painted on the blocks did not correspond with his idiosyncratic letter–color perceptions. Nabokov's mother readily understood because she too had colored letters as well as colored hearing,[16] demonstrating that while synesthesia is inherited, parent and child can have different types. In later years, Nabokov's son Dmitri would turn out to have grapheme → color and sound → color synesthesia.[17] Interestingly, Nabokov's wife Vera was also synesthetic, making it impossible to determine in retrospect from whom Dmitri inherited his synesthesia.

Susan Osborne escaped ridicule as a child because her father and sister also had grapheme → color synesthesia. To pass the time on automobile trips in the 1950s, they would play a game that involved calling out the corresponding colors of numerical sequences on roadside signs. For example, "Route 206" might make one of them sing out "sparkly blue, pale orange, and pink," whereas "Scranton 87 miles" might bring forth an assertion of "red and lime green." Of course, each player saw his or her own hues, and the fun part of the game was standing up for one's unique color combinations. Her mother, who was not synesthetic, never could make heads or tails of her family's strange pastime.

Given recent advances, synesthesia may be the first *perceptual* condition for which a gene can be discovered. It appears that synesthetic perception results from a heritable overinteraction between different areas of the brain. We explain these theories in chapter 9.

Nineteenth and early twentieth century reports emphasized that synesthesia appears to be common in childhood.[18] In 1883, the eminent psychologist G. Stanley Hall found that twenty-one of fifty-three children (40%) described musical instrument sounds as colored.[19] Thus, the overall impression is that synesthesia is several times more common in children than adults.[20] Because most synesthetes claim to remember having synesthesia far back in childhood, a plausible conclusion is that a proportion of childhood synesthetes lose their synesthesia as they grow up.

Consistent with this conjecture are a small number of our own case histories in which individuals claim to have lost their synesthesia around the time of puberty. For example, one man with a synesthetic brother recalls his own synesthetic childhood well but reports "by the time of my bar mitzvah it was gone." The time frame here is noteworthy, because puberty is a time of enormous physical change not only in outward appearance but also in brain reorganization. Juvenile brains are constantly changing and reorganizing themselves as they develop through childhood, but a big push comes during the concentrated secretion of sex hormones during puberty. Brain maturation then slows down dramatically until subsiding to a steady level in the early twenties.

In our experience, the number of people who claim to lose synesthesia is small compared to that of Riggs and Karwoski, who in 1934 wrote, "frequently synesthesia disappears in adolescence. . . ." As we do not have the benefit of having objectively evaluated their subjects, such evidence must be considered anecdotal. Shedding some light on this, a recent study[21] finds no difference in adults compared with children ages six to seven in the incidence of grapheme → color synesthesia. Either individuals with this kind of synesthesia tend not to lose it once it is established, or else if they do, the loss occurs before the age of six or seven.

Even less common in our experience than losing synesthesia around puberty are anecdotal reports of first gaining it. We know of two women with colored hearing who insist the trait was not present before thirteen years of age. Hinderk Emrich in Germany also cites anecdotal reports wherein synesthesia has either intensified or vanished at puberty.

Such observations have theoretical importance. A widely agreed upon idea among synesthesia researchers is that learning is involved in establishing synesthetic links at an early age. Once learned, however, the links appear to become fixed and remain stable for life. That at least appears to be true for the overwhelming majority of synesthetes. For example, Vladimir Nabokov jotted down Dmitri's letter–color pairs when the son was eight or ten. In his late thirties Dmitri chanced across his father's notebook and retested himself. The colors were all the same.[22] However, pubertal changes, no matter how seemingly rare with respect to synesthesia, suggest that the brains of synesthetic youngsters may actually be in a state of flux.

Early literature supports this notion. In 1917, Stanford psychologist David Starr Jordan, himself synesthetic, took note of the colored letters experienced by his eight-year-old synesthetic son.[23] Five years later, and without mention of the matter during intervening years, the father had the boy repeat his color list. Eleven of twenty-six (42%) of the boy's color–letter associations had changed somewhat.[24] A useful direction for future studies would therefore be to prospectively examine childhood color associations and see if they change over time during adolescence.

We can only speculate as to why synesthesia might be widespread in childhood but be lost with age as juvenile brains mature. Because it is invariable and overly inclusive as a mode of thinking, it may be that

Table 1.1
A possible cognitive continuum

Perceptual → similarities	Synesthetic → equivalences	Metaphoric → identities	Abstract language

Note. On a line of increasingly complex cognition, perceptions that are synesthetically "the same" fall between perceptions that are alike in some way (e.g., "bright" is "high") and those that are metaphoric (e.g., "loud color").

synesthesia is simply replaced by a more flexible mode of cognition, namely, abstract thought and language. As such, there may be a cognitive continuum beginning with perception, reaching to perceptual similarities, then to synesthesia, next to metaphor, and finally to abstract language (see table 1.1). In other words, synesthetic perception may be a normal stage of brain development that consciously persists in a minority of adults. We have more to say about synesthesia's fundamental importance to cognition in chapter 8.

What Synesthesia Isn't

There is confusion about the word "synesthesia" given that it has been used over a 300-year period to describe vastly different things ranging from poetry and metaphor to deliberately contrived mixed-media applications such as psychedelia, son et lumière, odorama, and even cross-disciplinary educational curricula. Therefore, we have to carefully separate those who use synesthesia as an intellectual idea of sensory fusion—artists such as Georgia O'Keeffe, who painted music,[25] or the composer Alexander Scriabin who included light organs[26] in his scores—from individuals with genuine perceptual synesthesia. This latter group of real synesthetes includes famous artists such as novelist Vladimir Nabokov, composers Olivier Messiaen and Amy Beach, and painters David Hockney and Wassily Kandinsky. Perhaps the observation that synesthesia appears to be more common among creative people fueled the dismissal that synesthetes were just being artistic rather than having a genuine sensory experience.

Synesthesia is not metaphoric language. If it were, we would expect the associations to change according to context rather than be stable as they are over a lifetime. We would further expect to find common expressions among different synesthetes instead of the highly idiosyncratic associations that are observed. Synesthesia is also not poetry, even though synesthetic-like associations are widely found in literary tropes. Poets commonly fuse different senses and employ cross-sensory adjectives to induce a compound aesthetic experience.

Another thing synesthesia is not is vivid imagination. What synesthetes see is not pictorial and elaborate but rather simple and elementary. For example, a common type of synesthesia is *colored hearing*, or the activation of color, shape, and movement by everyday environmental sounds, voices, and, especially, music. For synesthetes, these sounds trigger an experience something like fireworks: colored moving shapes that rise into existence then fade away. That is their perceptual reality. Whereas nonsynesthetes might picture a pastoral landscape while listening to Beethoven, for example, synesthetes will see colored lines or moving geometric shapes. Deni Simon describes her experiences this way:

When I listen to music, I see the shapes on an externalized area about 12 inches in front of my face and about one foot high onto which the music is visually projected. Sounds are most easily likened to oscilloscope configurations—lines moving in color, often metallic with height, width and, most importantly, depth. My favorite music has lines that extend horizontally beyond the "screen" area.

Deni's synesthesia is particularly vivid as evidenced by her perceptions' appearing to exist in a well-defined location. It is important to emphasize that synesthetes do not substitute or confuse one sense for another. That is, when seeing with their ears they do not *mistake* a sound for a sight. Rather they perceive both sensations simultaneously. In philosophers' terms, they have "extra" qualia, where qualia are defined as the subjective aspects of sensation such as redness, sweetness, or pain, for example.

Objective evidence that synesthesia is not imagination comes from brain scanning, which shows that brain activation patterns during synesthetic experience are not similar to those seen when subjects visualize in their mind's eye. Rather, the activations more closely match the patterns observed during actual perception. The idea that synesthetes have hyperactive imaginations and simply want to call attention to themselves is further contradicted by the fact that they do not buttonhole people to tell them how special they are. Rather, they often do the opposite, sharing their experiences only rarely either because they assume everyone has similar experiences or else because they have been ridiculed and subsequently keep their experiences private and hidden. Despite their reticence, synesthesia remains vivid and insuppressible throughout their lifetimes.

Studying Subjective Experience

Convention holds that *symptoms*—such as pain, dizziness, or forgetfulness—are subjective states "as told by" patients, whereas *signs*—such as inflammation, paralysis, or language errors—are objective, outwardly observable facts.

Historically, a long-standing problem with synesthesia was its lack of objectively observable signs. That is, reports of synesthesia relied entirely on subjective first-person statements by individuals claiming to have it. There is a long history in modern science of regarding self-reports and reference to mental states as unfit for investigation. As a methodology, introspection was considered scientifically unreliable because it was supposedly unverifiable. Such a stance against first-person accounts did not exist in the nineteenth century and earlier, when introspection was a popular experimental technique.

Not surprisingly, then, the late nineteenth century coincides with synesthesia's peak popularity as evidenced by the number of papers written on the topic (see figure 1.3). For example, during the fifty years between 1881 and 1931, seventy-four publications appeared, compared to only twenty-three in the fifty years between 1932 and 1982. The rise of behaviorism—a school of psychology that regarded behavioral observation (as opposed to investigation of conscious experience) as the correct approach to psychological inquiry—during the latter period is largely responsible for banishing mental states from scientific study. However, as the figure indicates, synesthesia is currently undergoing a renaissance.

Synesthesia was forgotten after its original heyday because science could not explain it. Simply put, psychology and neurobiology were premature disciplines at the time. Psychological theory relied on countless associations, and concepts of nervous tissue were paltry compared to today's understanding. What we consider modern neurology was still in its infancy, and neuroscience as we now practice it had not yet been invented. Just as

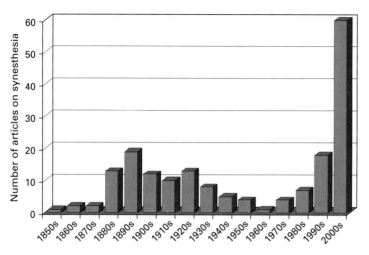

Figure 1.3
Peer-reviewed papers on synesthesia by decade from 1850 through 2006. There was considerable interest at the turn of the twentieth century, followed by a marked dropping off during the decades that behaviorism held sway as the dominant psychological paradigm. (Its height of popularity was between 1920 and 1940.) Increasing interest characterizes recent decades, indicating a second renaissance of synesthesia study.

concepts of how the brain was organized became recognizably modern, behaviorism appeared on the scene with such draconian restrictions against subjective experience that even acknowledging the existence of an inner life was taboo for a long time. In the eyes of the establishment, synesthesia was no longer a respectable topic of inquiry.

Believe it or not, relating conscious experience to brain function is also a relatively recent notion: neurology was overwhelmingly indifferent to mental life during the bulk of its history.[27] Rather, it concerned itself with movement, spinal reflexes, and other physical functions while leaving mental life to the realm of psychiatry and philosophy. Literature of the 1950s expressed considerable ambivalence about the status of the cerebral cortex in mental functioning,[28] an attitude finally swept away in the 1990s when the Decade of the Brain was ushered in.

How does science approach the distinction between a first-person understanding of experience and a third-person one that is supposedly objective? The lack of obvious agreement among synesthetes compounds the apparent difficulty. Indeed, the fact that two individuals with the same sensory pairings do not report similar, let alone identical, synesthetic responses has sometimes been cited as proof that synesthesia is not real. But this stems from a misunderstanding of both the neural mechanisms and the phenomenon itself and underscores the deeper issue of "Real to whom—to the questioner or the person who has it?"

While an inspiring knowledge base has arisen through physical investigation of the brain, it has proven harder to get a grip on mental life. Perhaps we can turn to the multiply talented nineteenth-century physician and physicist Gustav Fechner for guidance, because it is precisely physical absolutism that he meant to transcend. He articulated something that every nineteenth-century physiologist knew but that was gradually lost in the twentieth century's zeal to decipher the nervous system: namely, that a mental world exists. Then, as now, the question is how to do science in such an arena. In a fundamental sense, Fechner's psychophysics—relating physical stimuli to perceived sensation—has no substitute: no amount of brain imaging or analyzing nerve impulses can substitute for an introspective report. Even modern neuroimaging starts with the subject's state of mind.

How do we get an objective handle on a state of mind? Different kinds of scientists favor different kinds of data. Psychologists generally prefer to measure behavior, what people do, whereas biologically oriented types favor physical evidence, especially pictures of the brain. Synesthesia happily puts us smack in the middle of this dilemma by showing the necessity of using both first- and third-person accounts. It also shows how both kinds of observations are theory laden and how each kind biases conclusions.

An experimenter may observe, but only the subject has access to experience. Subjects, however, are prone to interpret their self-observations rather than give a straightforward description. It is not just that people ordinarily have no need to describe their conscious experience and thus do not pay attention to its subtle features. It is also that any description of experience contains implicit theoretical assumptions.[29] As they are based on what is called "folk psychology," meaning common sense, such assumptions often differ radically from those of the neuroscientist, who usually breaks experience into components and does care about its many subtle features.

For example, consider the experience of drug-induced visual hallucinations—a variety of "seeing things." Starting in the 1920s, the German psychologist Heinrich Klüver wanted to better understand them. He induced hallucinations with mescaline and quickly discovered that subjects were awed and overcome by the "indescribableness" of what they saw. They gave cosmic interpretations instead of straightforward descriptions, poetically embroidering their sensory experience. Once Klüver trained subjects to introspect carefully and hone descriptions down to essentials, he succeeded in identifying three categories of visual experience. We discuss one of these, the "form constants," in chapter 2 when we illustrate perceptual similarities that different synesthetes share.[30]

Just as subjects interpret or rationalize their experiences based on assumptions that are often very different from those of neuroscience, so too experimenters make assumptions when translating introspective reports into scientifically useful characterizations. Investigators do this either individually or as members of a collective neuroscience community. What they often fail to recognize, however, is that their assumptions and interpretations are just as theory laden as their subjects'. As a corollary, scientists cannot accept introspective reports literally, but subjects' biases can be addressed by training them to observe, and both subject and investigator bias can be minimized by using a script. Other sources of so-called objective data (e.g., brain metabolism as measured by scans) are also biased by investigators' assumptions. It follows that introspective reports—not accepted literally but properly guided and interpreted, and revised by investigators as necessary—are legitimate sources of data. Some years ago Richard cautioned, "though synesthetes are often dismissed as poetic, it is *we* who must be cautious about unjustifiably interpreting their comments."[31] Training ourselves is the flip side of training self-observers.

The opportunity to embroider is especially ripe in synesthesia because the experience is so ineffable. Because it is quite difficult to convey, subjects often resort to metaphor when describing "what it is like." For example, Michael Watson described the taste of spearmint as "cool glass columns." Was he being metaphoric, or verbally interpreting a sensory experience by couching tactile sensations in terms of images the investigator could understand? The distinction was teased out by asking him to focus on and describe the exact sensations he felt, rather than painting verbal pictures or "explaining" his experience. When given the spearmint stimulus, Michael rubbed his fingertips together and moved his hands through the air as if palpating an actual object:

I feel a round shape. There is a curvature behind which I can reach, and it's very, very smooth. So it must be made of marble or glass, because what I'm feeling is this incredible satiny smoothness. There are no ripples, no little surface indentations, so it must be glass because if it were marble I'd be able to feel the roughness of the stone or the pits in the surface. It's also very cool so it has to be some sort of glass or stone material because of the temperature. What is so wonderful is the absolute smoothness of it. I can run my hand up and down, but I can't feel where the top ends. I feel that it must go on up forever. So the only thing I can explain this feeling as is that it's like a tall, smooth column made of glass . . . there is a funny sort of feeling of being able to reach my hand into this area. It's very, very pleasant.[32]

It quickly became evident that Michael experienced elementary tactile qualities that he sensed as identical over time if given the same stimulus. Questioning other synesthetes about their experience and retesting them at later dates subsequently confirmed this stability, leading to one of the five diagnostic criteria for the trait, namely, that synesthetic percepts are generic and consistent. "Generic" means that what is experienced is not complex or pictorial but elementary—blobs, lattices, cold, rough, sour, zigzag, or geometrically simple. "Consistent" means the experiences are largely (but not totally) invariable through time. Teasing out these phenomenal features therefore contributed to synesthesia's classification as a real condition.

Had Richard accepted Michael Watson's descriptions at face value or as metaphoric, rather than probing to tease out the sensory qualities Michael struggled to convey, synesthesia might have remained relegated to a mere curiosity rather than being recognized today as a striking anomaly with important implications for concepts of how the brain is connected. Rather than second-guessing experiential reports, the focus of those who currently explore the phenomenon has been on trying to understand synesthetes' reports and seeking behavioral correlates of their subjective claims. This requires interplay between first-person and third-person accounts. A dialogue or structured questioning between clinician and subject constitutes a second-person relation between shared knowledge about experience. This kind of feedback sometimes leads to further self-observations from synesthetes and the subsequent invention of additional third-person tests by the investigator.

Theoretical constructs constituting investigator biases, though evidently derived from objective third-person observation, nonetheless have historically shortchanged first-person accounts. For example, Richard first became aware of the mismatch between subject and examiner during his ophthalmologic training in the 1970s. He was struck by how often patients spoke of "seeing things" they could describe in detail sufficiently nuanced so as to sound similar to "things" other patients saw, yet the medical exam revealed nothing amiss despite using plenty of special equipment. Faculty and trainees assumed that every optical symptom had a physical correlate they could observe through their lenses. The conceit never had to be taught explicitly because it was just taken for granted, deeply embedded.

Looking back, that level of confidence is breathtaking, but many assumptions investigators make appear so "obvious" that their bias is never questioned. Patients did not want to hear "I don't see anything wrong with your eyes," because it left their questions unanswered, and their persistent asking left physicians frustrated. Each side had different expectations. Perhaps Richard's sensitivity to the mismatch between subject–observer expectations left him open to Michael Watson's experience instead of dismissing it as impossible, or Michael as crazy, as his medical peers automatically had. Investigator bias toward or against something can therefore influence the very kind of data one takes up as interesting.

A Different Texture of Reality

Imaging being able to see into the ultraviolet part of the electromagnetic spectrum the way birds and bees do—or into the infrared or X-ray range the way machines can. You would literally perceive the world differently from other individuals and have a different texture of reality than other people. This is similar to the situation of a synesthete, who experiences the world more richly.

Often nonsynesthetes imagine that sensing nonpresent colors, textures, and configurations must somehow be a burden. "Doesn't it drive them crazy having to cope with all the extra bits?" they ask. But this is no different than a blind person telling you, "Oh you poor thing. Everywhere you look you're always seeing something. Doesn't it drive you crazy always having to see everything?" Of course not, because seeing is normal to us and constitutes what we accept as reality. Synesthetes simply have a different texture of reality.

This fact brings us back to a central point—namely, that reality is much more subjective than most people suppose. Far from being objectively fixed "out there" in the physical world[33] and passively received by the brain, reality is actively constructed by individual brains that uniquely filter what hits the outside senses.

In this light, synesthesia catalyzes a paradigm shift by highlighting the dramatic differences in how individuals objectively see the world. It illuminates a broad swath of the mind and forces fundamental rethinking about how brains are organized. Over a quarter of a century ago, neurologists insisted that synesthesia could not be real because it contradicted accepted theory; today, synesthesia's undeniable reality is forcing theory to change. Understanding this strange and delightful way of perceiving can help us understand how all our brains operate.

2 A Kaleidoscopic World

An astounding two-thirds of our approximately 25,000 genes are expressed in the brain. Because genes can be transcribed in a variety of ways and further influenced by external forces (called "epigenetic"), synesthesia occurs in an amazing array of combinations. Having one type of synesthesia gives you a 50% chance of having a second or third type, meaning that the gene may be active in a number of different brain areas simultaneously. We call individuals with multiple kinds of synesthesia "polymodal."

Table 2.1 lists how relatively common different types of synesthesia are. The tabulation of frequencies was compiled by Sean Day, an American linguist who is himself synesthetic. He also moderates an online community called The Synesthesia List that brings together synesthetes, researchers, and curious individuals.[1] Because the table's data rely on self-identified synesthetes rather than random sampling, it is biased in the same way surveys are as we discussed in the previous chapter. For example, Sean Day's sampling found a female-to-male sex ratio of 2.5 to 1 and found that grapheme → color is the most common manifestation—neither of which has held to be true after Julia Simner and her colleagues studied random slices of the population.[2] They showed a roughly equal number of women and men and that the most common synesthesia type is colored days of the week. Still, the table is useful so long as we keep this caveat in mind.

On inspecting table 2.1 two observations stand out. First, some combinations are far more common than others. Look in retrospect at how rare Michael Watson's tasting shapes was: only half a percent. It is equally obvious that color is overwhelmingly the most common synesthetic response, whereas graphemes, time units (such as days of the week), and sounds are the most common synesthetic triggers. Second, notice that synesthesia usually goes only in one direction, such as sound → vision but

Table 2.1
Relative frequency of different types of synesthesia

Type	Frequency (%)
Graphemes → colors	66.50
Time units → colors	22.80
Musical sounds → colors	18.50
General sounds → colors	14.50
Phonemes → colors	9.90
Musical notes → colors	9.60
Smells → colors	6.80
Tastes → colors	6.60
Sound → tastes	6.20
Pain → colors	5.80
Personalities → colors	5.50
Touch → colors	4.00
Sound → touch	4.00
Temperatures → colors	2.40
Vision → tastes	2.10
Sounds → smells	1.80
Vision → sounds	1.50
Orgasm → colors	1.00
Emotions → colors	1.00
Vision → smells	1.00
Vision → touch	1.00
Smells → touch	0.60
Touch → tastes	0.60
Smells → sounds	0.50
Sounds → kinetics	0.50
Sound → temperatures	0.50
Tastes → touch	0.50
Kinetics → sounds	0.40
Personalities → smells	0.40
Touch → sounds	0.40
Touch → smell	0.30
Vision → temperatures	0.30
Musical notes → tastes	0.10
Personalities → touch	0.10
Smells → tastes	0.10
Smells → temperatures	0.10
Tastes → sounds	0.10
Tastes → temperatures	0.10
Temperatures → sounds	0.10
Touch → temperatures	0.10

Note. Comparative frequencies of different kinds of synesthesia. Data are based on Sean Day's tabulation of 738 self-reported cases from a nonrandom sample. In Day's sample, 72% were female and 28% were male. (Data are reproduced with permission from http://home.comcast.net/~sean.day/html/types.htm.)

hardly ever vision → sound. This should have been another clue to skeptics that synesthesia was real, because neural circuits are known to be directional whereas made-up stories would have no reason to be. Lastly, while we usually refer to synesthesia as cross-sensory pairings, the table makes clear that a wide variety of nonsensory cognitive traits (such as personality) can couple with various aspects of sensation. This makes the phenomenon all the more intriguing.

Recently, David's laboratory rigorously verified grapheme → color synesthesia in 1,067 subjects (chapter 3 gives details of the testing). Additional types of synesthesia reported by members of this group appear in figure 2.1. Weekday → color and month → color synesthesias, listed as time units → colors in table 2.1, are the most common co-occurring synesthesias. This is not surprising given the high prevalence of these forms in general. Musical sounds → color and spatial sequences are tied for being the next most common types, while almost half of grapheme → color synesthetes seem to have one or both of these. Note that spatial sequence synesthesia, discussed below and in chapter 5, are not included in Day's survey. Also note that with the exception of spatial sequence synesthesia, the other most common forms of co-occurring synesthesia all involve color, while those involving touch, taste, sound, and smell are comparatively rare. Forty-five of the 203 synesthetes (22%) report having no other

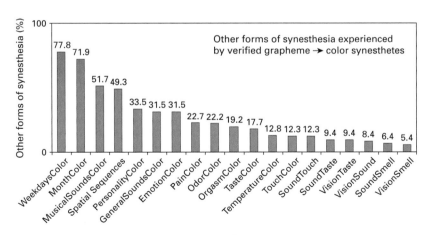

Figure 2.1

Other types of synesthesia reported by 1,067 verified grapheme → color synesthetes. Data are from David Eagleman's Synesthesia Battery.

form of synesthesia. Which synesthesias tend to co-occur, and why, is revisited in chapters 9 and 10.

Common Synesthesia Types

To introduce the broad array of synesthesia types, we will begin by briefly elaborating five common types of synesthesia: number forms, colored letters (both written and spoken), tasted words, colored hearing, and the personification of letters and numbers.

Number Forms

The coupling of color, perspective, and spatial configuration with concepts involving sequence, or ordinality, was noted more than a century ago.[3] To those who "see" number forms (also called "spatial sequence synesthesia"), numbers or other ordered concepts typically lie on a path that twists and zigzags in all manner of shapes. Often it loops around and encircles the body as it executes a variety of angles, bends, and curves (see figure 2.2). Such patterns differ in different persons. Figure 2.3 shows several simple forms that Galton illustrated in 1883:[4]

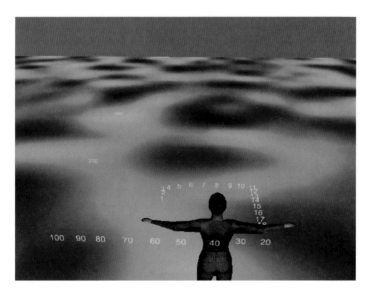

Figure 2.2
A clockwise number form with right angles that encircles the body.

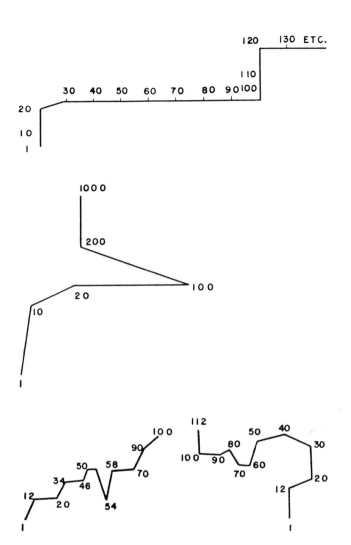

Figure 2.3
Number forms reported by Sir Francis Galton, 1883.

The drawings, however, fail in giving the idea of the apparent size to those who see them; they usually occupy a wider range than the mental eye can take in at a single glance, and compel it to wander. Sometimes they are nearly panoramic.

These forms . . . are stated in all cases to have been in existence, so far as the earlier numbers in the Form are concerned, as long back as the memory extends; they come "into view quite independently" of the will, and their shape and position . . . are nearly invariable.[5]

Like other kinds of synesthetic perception, number forms have a dynamic quality as well—recall the student who complained to her math teacher of digits that "kept going up to their places." In the classroom, Nobel physicist Richard Feynman saw colored equations floating in space in front of him:[6] "As I'm talking, I see vague pictures of Bessel functions . . . with light tan j's, slightly violet-bluish n's and dark brown x's flying around. And I wonder what the hell it must look like to students."

Synesthetes possessing number forms express amazement that not everyone "sees" numbers as they do or that anyone finds such spatial configurations odd. As grapheme synesthete Marti Pike puts it, "It never occurred to me that it might be unnatural to visualize the whole alphabet or numbers." People like her describe the forms vividly, matter-of-factly and in the present tense.

Number forms do not indicate high mathematical ability or deficiency, nor do they seem to be correlated with any specific intellectual talent or mental dullness. However, the automatic positioning of integers can interfere with more complex mathematics, such as algebra or calculus, as in the wayward numbers of the student cited above. "For one thing," notes Marti, "they are not evenly or consistently spaced. There is also some fluid or jelly-like movement to them." Magnitude can also be problematic because, for Marti, 6 is physically the highest number in her visual representation. Therefore, 6 and numbers containing 6 have the highest degree of magnitude. In other words it does not "make sense" to her that 11 is "larger than" 6 or that 234 is "greater than" 66. To her, magnitude comparisons are literally physical: when she thinks of someone older than herself, she looks "up," whereas younger people are seen not "down" but "in back" of herself.

Alphabet, time, and calendar forms are especially common. In the latter, it is usual for the months to take up unequal space, and the topmost month is by no means necessarily January (see figure 2.4).

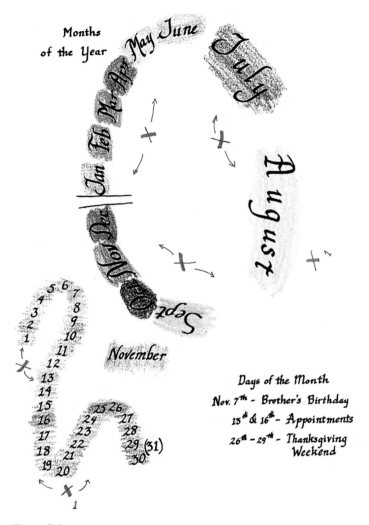

Figure 2.4

Calendar and month forms for Marti Pike. Note that June is topmost and that July, August, and September take up more space than other months. Brown-colored November gives way to a brown serpentine form for days of that month. "Highlighted" days mark appointments, birthdays, and special occasions, aiding her memory.

Perhaps in number forms, more than in other kinds of synesthesia, one sees how focused attention modifies what is perceived. Seers read their panoramas like a map, zooming in and out and changing perspective within the array. They can look right or left, up to or down on it, as if embedded in a virtual reality display or, as Marti puts it, "a flexible moving 3-dimension." She likens what she privately calls her "memory maps" to a geographical map that allows you to take in an overall view without much detail and then zoom in on a specific enlarged region of interest. When moving about like this, the "illumination" changes so that the elements at the center of focus are spotlighted or highlighted in some way. Another synesthete (HC) speaks of "a window" in which she views six or seven colors at a time while scanning her number form. Although synesthetes can change their viewing perspective, the relationship between elements in the form remains constant.

Though she painstakingly rendered her drawings in colored pencil, Marti insists they fail to capture their panoramic and three-dimensional aspects. Transparency and other subtleties are also left out. She sees her alphabet and numeral forms against a black background (see figures 2.5 and 2.6). The tiny red x's in the figures denote one of many possible points of view from which she can survey the forms. "Looking back at my illustrations they look ridiculous, even to me . . . I'm not used to seeing the map in one dimension or so finite with such limiting borders. The maps are larger than my visual range, like looking at the horizon."

Colored Graphemes

We can use Marti Pike as an example of someone who senses colored graphemes and take advantage of her experience to say something about how learning and synesthesia intertwine. Obviously, letters and numbers are not innate. At some point children must learn these cultural artifacts. When and how, for example, did Marti learn that "A" is red? And once learned, how did her genes act in her developing brain to make the "A = red" association not only permanent but also vividly perceptual? The short answer is that we do not yet know the mechanics in detail. Scientifically, we are still at the descriptive stage of how synesthesia and learning interact, but any general theory of synesthesia must account for both neural development and acquired experience (i.e., learning one's own particular language). Furthermore, learned graphemes other than

Figure 2.5
Marti Pike's alphabet form. Letters A and Z feel to her as if they are anchored in a black background, while the loop floats free in three dimensions. Letters from J to S are spotlighted. Red x's indicate the ability to shift viewpoints in the panorama (e.g., look for the x's in the alphabet shape near the letters D, S, M, and V). (From Cytowic [2002], with permission.)

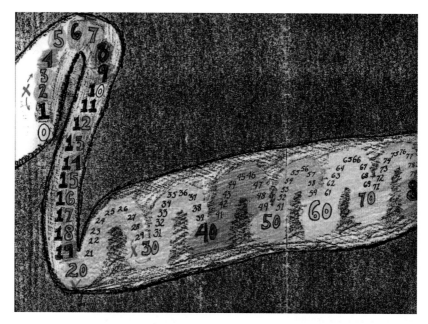

Figure 2.6
Marti Pike's numeral form. 1 to 20 are seen singly, and 20 to 100 in groups of 10. After 100, the pattern repeats with a 1 in front of it. She can view 100 numbers at a time (e.g., 50–150). Note the upside-down clock (nos. 1–12), which she had to unlearn when learning to tell time. Compare this to her decade form (figure 5.6). (From Cytowic [2002], with permission.)

letters and numerals—such as punctuation marks, Braille, and musical notation—can also become bound to color. Although the final answer is not yet in, neurobiologically, in chapter 9 we will outline the contours of a theory.

Colored graphemes are dealt with at length in the next chapter. The point we wish to stress in this brief overview is that alphabet colors are typically determined by graphemes rather than phonemes. That is, how something is spelled matters more than how it sounds. Thus, "fish" and "photo" look different even though they sound similar. "Cathie" may look synesthetically different from its homonym "Kathy," and "Brown" different from "Browne" (see figure 2.7). Unsurprisingly, grapheme-based synesthetes like Marti say that the colors help them remember names, particularly those with less common spellings. Numbers also have their

Figure 2.7
In grapheme-based synesthesia, homonyms look different. Often the first letter "shades" the entire word, whereas vowels lighten or darken it.

own color that is independent of the letters that constitute the name of the number. Furthermore, the input sense is irrelevant: color appears whether the grapheme is read, heard, or merely thought of.

Very often the first letter of a word shades its overall color, whereas vowel colors lighten or darken it. For idiosyncratic reasons, synesthetes commonly say that word shading is "important," and it would be weird if words lacked this property. In Marti's case, A and I darken words, E and O lighten them, and U is neutral. As can be seen in figure 2.7, "TIN is darker than TAN, TAN darker than TEN (and of course '10' looks different), TEN darker than TUN. TUN (a large cask of wine) is a color in between TIN and TAN or TAN and TEN. If E is added to TUN to make TUNE, the word becomes very much lighter." Next in importance to Marti are letters that she perceives as "spatially big." That is, B has more color than C, and K more color than L. Marti likens the phenomenon of word shading to looking at threads in a tapestry:

The letter colors appear like fabric; pull the individual threads apart and you can see the various colors. Woven together, they become one, i.e., predominantly green (T), but influenced by other "strands." For example, say you are looking at a rug. One of two strands out of 5 will be brown, the rest green. Look at the fibers running through the strands. Some are red (A–TAN), or perhaps yellow (E–TEN), or gray (U–TUN), or maybe black (I–TIN). Standing on it, you say that the carpet is green; sitting on it you might notice the brown, and examining it closely, you may see the red, yellow, etc.

This is another example of how focused attention influences the synesthetic percept. Shifting the focus of attention can literally change what is perceived. For example, the synesthetic color of the digit 5 in figure 2.8 changes depending on whether a synesthete attends to the large 5 or instead to the smaller constituent 2s.

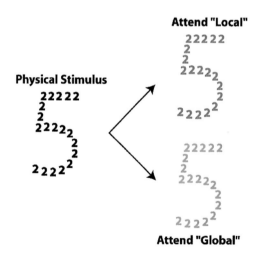

Figure 2.8
Shifting attention literally shifts the synesthetic perception. (From Palmeri et al. [2002], with permission.)

The Phoneme Minority

Research shows that by six months of age, most children recognize the basic sounds (phonemes) of their native language. When children are around four to five years of age, phonemes give way to graphemes as children learn the alphabet and words. It is therefore most peculiar that synesthetic color is overwhelmingly attached to graphemes rather than phonemes. For reasons we do not yet know, synesthesia performs its coupling action during this later stage of brain and language development during mid-childhood. At this age, a child's brain is still reorganizing itself dramatically.

Marti Pike's earliest recollection of the alphabet dates from age four, a typical age for learning letters,[7] when she recalls spelling M–I–L–K from a milk carton. When her grandmother asked her how to spell TEA, Marti remembers that the letter "T" was in the "dark part" of the alphabet that she "could not yet see." Her alphabet was only beginning to take shape and was still "clouded in darkness." She was an early reader, learning to read before starting school at age five, by which time she remembers her alphabet pattern being firmly fixed. The days of the week appeared shortly thereafter, and her form for the months was established by approximately age seven.

The role of learning raises the issue of critical periods. The scientific concept differs from the popular fad of neonatal enrichment epitomized by playing Mozart to the womb. The latter wrongly claims there are only a few critical times when children must be exposed to certain stimuli without which they will miss crucial developmental steps.[8] Neurologically speaking, it is never too late for Mozart. Rather, critical periods exist during the normal development of specific functions such as hearing or vision when an infant must have sensory or motor experience—otherwise a window closes and the opportunity is lost forever. In one famous experiment a newborn kitten's eye was temporarily sewn shut. After many weeks the sutures were removed, but the mature cat never developed proper vision through that eye because activity from that eye to the brain was lacking. In other words, the environment (the visual world) was prevented from reaching the brain.

The attachment of synesthetic color to graphemes rather than phonemes raises the question of whether the synesthesia gene expresses itself during a critical period of childhood brain development, and if so, how.

If synesthesia were entirely innate, it would be plausible to expect language-based synesthesia to always be sound based, or phonemic, given that phoneme perception is present at birth and grapheme perception only from ages three to four years at the earliest. Yet in about 10% of synesthetes, phonemes do trigger color. "Fireman," "pheasant," "off," and "enough" might all be blue words because each contains a prominent /f/ *sound*. There is also a small subset of people who sense color only in response to hearing the spoken word. Six such individuals have been studied by positron-emission tomography (PET) scanning.[9] It is odd that these individuals experience synesthesia only upon hearing words rather than reading them, because adult readers automatically connect the *sound* of a word with its *visual* appearance (in technical terms, the orthographic lexicon is activated).[10]

Evidently there are also in-between cases of speaking, seeing, or hearing a trigger. Listen to Vladimir Nabokov describe his "fine case of colored hearing":

Perhaps "hearing" is not quite accurate, since the colour sensation seems to be produced by the very act of *orally forming a given letter* while I imagine its outline. The long *a* of the English alphabet (and it is this alphabet I have in mind unless otherwise stated) has for me the tint of weathered wood, but a French *a* evokes polished ebony. This black group also includes hard *g* (vulcanized rubber) and *r* (a sooty rag being ripped). Oatmeal *n*, noodle-limp *l*, and the ivory-backed hand mirror of *o* take care of the whites[11] [*italics added*].

How do we account for the seemingly anomalous persistence of colored phonemes in some synesthetes? Perhaps grapheme-based synesthesia begins as phoneme synesthesia and then undergoes a conceptual reorganization with phoneme-to-grapheme conversion rules during the acquisition of literacy. There is no a priori reason why young synesthetic brains should not undergo the same developmental reorganizing pressures as everyone else's. Prospective testing of synesthetic children would be necessary to prove such a conversion hypothesis, however, and that would be hard but not impossible to do. Sound-based synesthetes would then be seen as individuals whose synesthesia gene got expressed at an earlier than normal age before phoneme-to-grapheme conversion took place, thus permanently linking their colors to sound rather than written letter forms. This framework supposes that in most instances the synesthesia gene does not express itself until mid-childhood.

Tasted Words

Whereas graphemes generally determine color, phonemes tend to evoke synesthetic tastes. Presumably in this larger group the synesthesia gene is acting at an age when children are learning the names of what they eat. Table 2.2 lists the customary age of various learning milestones and the synesthesia type that might possibly result at each stage.

Briefly, synesthetic tastes are located in the mouth and occur for both spoken and written words. Tastes are usually specific rather than basic tastes such as salty or bitter. For Mathew Mousatkis, for example, "Steve" tastes of poached eggs. In this instance temperature and texture are implied, but often they are explicit: for James Wannerton, "jail" tastes of bacon, cold and hard.

Not every word elicits a taste. Common words are more likely to elicit tastes than infrequent ones, and actual words (technically called "lexemes") are far more likely to evoke tastes than made-up nonwords such as "polt" or "tweal."[12]

Like grapheme synesthetes who take the trouble to precisely describe their color experiences (e.g., "very pale violet with darker flecks"), gustatory synesthetes also describe their tastes exactly ("profit" = oranges, unripe, pithy). For a given word, tastes are likewise consistent over time. Unlike grapheme synesthesia, however, the initial letter has no special status. That is, words sharing the same first letter tend not to produce the same taste. Also, the taste of a given word is not readily predicted by the taste of its individual constituent letters. There are, however, many sound (phonological) and meaning (semantic) relationships between the triggering word and the food name experienced as the synesthetic taste (e.g., "rice" tastes like rice, and "onion" like onion).

Words containing similar patterns of phoneme sounds (rather than graphemes) tend to elicit the same taste (e.g., "television" and "Kelly" both elicit the taste of jelly). Furthermore, the critical phonemes tend to be contained in the food name reported as the synesthetic taste (e.g., "Barbara" = rhubarb). Subsequently, words that share either *sound* ("April" = apricots) or *meaning* ("baby" = jelly babies) with food words may acquire the corresponding synesthetic taste.

In pointing out that phonemes triggering tastes tend to be present in the word for that taste—for example, "dogma" tasting like hot dogs, and "super" like tomato soup—we again note the need for a better understanding of the interplay between nature and nurture. Although there is

Table 2.2
Age of acquisition for cognitive traits and synesthesia types

Age of acquisition of cognitive traits

6 mos	1 yr (12 mos)	1.5 yr (18 mos)	2 yr (24–30 mos)	2.5–3 yr (30–36 mos)	4 yr (36–48 mos)	5 yr (48–60 mos)	5–7 yr (60–84 mos)
Phoneme perception	Word fragments, words	5–20 word vocabulary	Oral speech: phrases, object names, 1° colors	3-word sentences; 2° colors; self-gender comprehension; mastery of food names; naming 1–10[a]; some alphabet[b]	Identifies with same-sex parent; cooperative play; visual-tactile cross-modal associations	Prefers gender appropriate games; attributes characteristics to others; dow, moy, clock reading[c]	Reading, writing
6–12 mos		18–24 mos		36–42 mos		48 mos	
Food preferences obvious		Food name use begins		1–10 in order; numbers > 10; alphabet (42–48: knows out of sequence)[d]		Early readers	

Possible age of acquired synesthesia

6 mos	1 yr (12 mos)	1.5 yr (18 mos)	2 yr (24–30 mos)	2.5–3 yr (30–36 mos)	4 yr (36–48 mos)	5 yr (48–60 mos)	5–7 yr (60–84 mos)
	18–30: Tasted phonemes		30–36: Colored phonemes, colored tastes	34–48: Colored graphemes		30–60: Emotionally mediated	
				34–60: Number forms		34–40: Lexical (spoken words)[e]	
						36–60: Gendered/personified graphemes	

Note. mos = months. dow = knows days of week. moy = knows months of year.

[a]But does not yet know digits 1–10 out of order. [b]At 30 months, some kids know the alphabet by song. [c]Many number forms contain clock shapes for the digits 1–12. [d]Between 42 and 48 months recognizes letters out of order. [e][spoken words → color] could possibly occur much earlier; young speakers recognize the lexicality of words/nonwords before learning to write and read. It could occur shortly after colored graphemes but before phoneme-to-grapheme conversion. One question is whether word → color and grapheme → color synesthetes are early readers.

an innate component to taste synesthesia, it is heavily influenced by learned vocabulary and conceptual knowledge. Chapter 6 is devoted to synesthetic taste and smell in detail. We discuss there why lexicality (word vs. nonword) and word meaning are important in this kind of synesthesia.

Colored Hearing

The term "colored hearing" refers to the elicitation of color, shape, and movement by sound. Triggers include everyday environmental sounds such as dog-barks, clattering dishes, voices, and especially music. We think the dynamic analogy to fireworks is apt because the colored shapes appear, move around a bit, then fade only to be replaced by other colored shapes (photisms) in a kaleidoscopic montage so long as the varying sound stimulus continues.

As for voice stimuli, photisms are determined purely by acoustic properties rather than by any semantic meaning of what is said. As Rebecca Price notes:

One of the things I love about my husband are the colors of his voice and his laugh. It's a wonderful golden brown, like crisp, buttery toast, which sounds very odd, I know, but it is very real.[13]

What a dog's bark looks like to Henry Gilbert depends on the dog. The deep woof of a large dog such as a German shepherd is black and gray like pepper, whereas a small yipping dog makes a sound of white circles.

Mary Lou Luff sees colored shapes projected above and in front of her when listening to music:

If I am tired at the end of the day the shapes seem very near . . . shiny white isosceles triangles, like long sharp pieces of broken glass. Blue is a sharper color and has lines and angles, green has curves, soft balls, and discs. It is uncomfortable to sit still [possibly indicating a secondary (sound → movement) synesthesia]. I feel the space above my eyes is a big screen where this scene is playing.[14]

A number of acquired synesthesias involve sound-to-sight coupling. They can occur in sensory deprivation, drugged states, and various forms of adult-onset blindness. We discuss these in chapter 9.

Personified Graphemes

Not only do Megan Timberlake's letters, numbers, and punctuation marks have color, but they also "additionally . . . seem to carry personality and gender, the same way they carry color. They are 'real characters!' (pun intended)." When first examined in 1985, she appeared to have an unheard-of type of synesthesia until a literature search turned up identical cases of personification reported in the 1890s.[15] The novelist Vladimir Nabokov also had personified graphemes.

Table 2.3 lists the colors, genders, and personalities of Megan's graphemes. Her associations are consistent, showing no variation over five months and again when retested at two years. It is interesting to note that the letters of her initials, M ("my favorite color for 'my' letter") and T, are both masculine.

For Megan, multiple-digit numbers and groups of letters forming words contain the colors of their constituent elements. In this respect she differs from Marti Pike for whom the first letter determines a word's overall color. Numbers 11 to 19 retain the gender of the final digit, but there is a blending of personality characteristics of the composite digits. Beginning with 20, however, the genders are determined by the first digit alone. For example, 20 is male whereas 40 is female. The 40s, 400s and 4,000s are all female. Similarly *words* take on the gender of the initial letter, but not the color or any other attribute. We will turn to personified graphemes in more depth in the next chapter.

Unusual Types of Synesthesia

Table 2.1 listed a number of infrequently encountered sensory combinations. We elaborate briefly on a few of the more interesting and unusual types of synesthesia we have encountered.

Audio–Motor Synesthesia

One of the most unusual and rarest types of synesthesia recorded is sound → movement, called "audio–motor" by the physician who described a single case of it in 1966.[16] An adolescent boy assumed different postures in response to the sounds of different words. The boy claimed that both English and nonsense words compelled him to make certain movements, which he demonstrated by striking various poses. The physician planned

Table 2.3

Color, gender, and personalities of Megan Timberlake's graphemes

Alphabet	
A	Bright-to-medium yellow; female, very feminine (always in dresses)
B	Orangey–beige, medium tonality; female; sturdy in character
C	Sky blue or slightly deeper; male; a touch impetuous, but generally dependable
D	Deep charcoal, nearly black; male; dashing, a bit of a joker
E	Light lavender; male; a soft-spoken type
F	Brownish–woody–colored, medium tonality; male; perpetually youthful, easygoing, casual character
G	Slightly lighter than medium purple; male; rugged good looks and credible
H	Orange but toned down (lighter than B); female; of a more formidable figure than A, but just as feminine
I	White, with "dirty edges"; male; a bit of a worrier at times, although easygoing, sincere
J	Violet–red–violet (purplish with a reddish cast); male; appearing jocular, but with strength of character
K	Yellow–beige (more to beige than yellow); female; quiet, responsible
L	Beige/tan/kaki: a specific shade; male; handsome, easygoing, adult with a thickened figure
M	Blue–violet (my favorite color for "my" letter); male; secretive, powerful, handsome
N	Medium-to-deep green, but the lightest of the green characters (which include T, Z, and 2); male; youthful, handsome, mediation type
O	Clear; the color of (clean) water; female; quiet, warm, reliable, of balanced character
P	Orangey, browner than B; female; busy, fun, sisterly
Q	Cranberry; female; elegant, nontalkative, more earthy than the playing-card depiction of Q (as in "queen")
R	Red; female; all-American woman, outgoing, active
S	Pastel yellow (lighter and less yellow than A); female; independent but good at partnering: mature
T	Forest green; male; quite masculine, and quiet, gentle, mature, responsible, slim build, handsome, good in relationships
U	Soft pink, rosy but not "pinkey-pink"; female; of rounded, not slim, figure; sweet, hardworking, quiet
V	Yellow beige but subdued (more beige than A; deeper than S; more yellow than L or K); female; very feminine, unflauntingly sexy; sophisticated

Table 2.3
(continued)

Alphabet	
W	Medium gray; "clean" gray; male; open-minded, seeming older than other characters; good-looking, friendly
X	Cheddar cheese color but deeper; androgynous; easygoing and fun loving; of balanced character; sometimes cheerful, sometimes worried
Y	Medium-to-deep gray; male; effeminate, attractive, responsible
Z	Deep forest green; male; dashing and very handsome because of it; mature but still playful; reliable

Numbers	
1	White; male; quiet character; youthful appearance but serious in character
2	Green with an almost bluish tinge; masculine; good-looking, somewhat outgoing, laughs easily, kind
3	Lighter lavender than E; masculine; teddy-bearish character, can seem gruff but isn't
4	Very yellow: deeper than A; female, feminine, playful but not flirtatious; sisterly
5	Blue, deeper than sky blue but similar to C; male; a bit of a worrier but not without self-confidence; mature
6	Pink, soft pink like U but slightly deeper; masculine; youthful, quiet, smart
7	Dark grainy gray, with an almost beigey sense to it; male; playful, and impressively handsome
8	Orangey–beige like H, but more beige and somewhat lighter in tone; masculine; cheerful, modest, firm figured
9	Dark, almost purplish, charcoal gray; male; a bit of a "wise guy" but nevertheless dependable; with a serious and powerful side
0	Zero is the same color as the letter O, that is, transparent, like the color of water; female; more subdued than O as if the less dominant twin
10	(and powers of 10) Although like all other combinations of letters and numbers 10 retains the individual colors of its component numbers, it also has a redness that lies under the white of 1 and the transparency of 0; the powers of 10, 100, 1,000, and so forth are all female, and increasingly formidable (as 10 is formidable); they are all friendly, matriarchal, and independent
11–19	Retain the gender of the *final* digit, and a blend of the characters of the composite digits

Table 2.3
(continued)

Punctuation marks

?	Cranberry–purple; masculine; respectable, serious, mannerly
,	Dark–gray; male; a bit irreverent
. : and !	Blackish, charcoal gray; masculine
"	Beige; masculine
'	Black
+	Yellowish; female; subdued character
#	Beige; male, hardworking

Groups of similar colors

A, S, V, 4, @, $, +	I, 1
B, H, P, 8	J, Q, ?
C, 5	K, X, #, +
D, (W), Y, 7	N, T, Z, 2
E, 3	O, 0
F, L, "	(R, 10)
G, M	U, 6

to retest the boy later on without warning to convince himself of this sound-to-movement synesthesia. Ten years later when the doctor read the same word list aloud, the boy assumed without hesitation identical postures to those of a decade earlier.

Although no further cases appear in the literature, the mapping of sound to movement is not as strange as it may seem at first glance. Just as sight and sound are closely integrated senses in everyone (which is why at the movies we hear the sound as coming from the actor's moving lips rather than from the surrounding speakers), so too are sound and movement. Think of dance, where body rhythms map to sound rhythms both visually and kinetically. As we discuss in chapter 8, such tight binding between two senses is not synesthetic, although it does help us to understand similarities between synesthetes and nonsynesthetes and to clarify why common metaphors make sense.

Geometric Pain

According to Sean Day's tabulation (shown in table 2.1) pain causes color in 5.8% of cases. He lists no instances in which pain is experienced as a

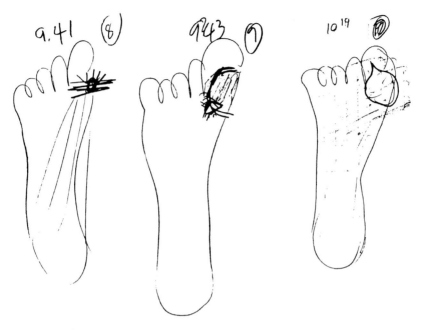

Figure 2.9
Time sequence of shaped pain following foot surgery in Rita Bush. The shape changes over time just as the experience of pain is neither constant nor monotonous. Note the simple geometry. (From Cytowic [2002], with permission.)

shape, although we have come upon such a case. Rita Bush describes her experience in this way:

As far back as I can remember I have felt pain in shapes, though it seems to me anyone could do so easily. Often, but not always, I hear voices (particularly singing voices) in shapes; I have felt at times I could draw or paint a song.[17]

Her shapes are felt, rather than seen, on the skin surface. They are never intricate but rather simple blobs, grids, cross-hatchings, and geometric forms (see figure 2.9). As her drawing illustrates, the shapes are dynamic and change over time.

Blindsight

MD was born with severely restricted vision. By adolescence she could only distinguish light from dark, and by the time she entered college she was completely blind. Nonetheless, for as long as she can remember, each letter of the Braille alphabet has had an unchanging color. "Some letters seem to emit light," she says. Because she did have some useful vision as a child she learned the Roman alphabet. But, she says, "If I insist on seeing the same [Braille] word in [Roman] letters . . . then the letters have the same colors as their Braille representations. They are often somewhat odd shades or hues that I would not consciously choose."

This last comment suggests that her V4 color area is being cross-activated by the concept of both Braille and Roman graphemes. Additionally, MD is polymodal. She "sees" the white keys on a piano and those on a QWERTY keyboard as colored. She also has number forms for numerals, months of the year, and days of the week. Physical streets (not their names) are colored, as are the U.S. states. Geographical maps are obviously colored, and she also has [music → color, shape] synesthesia.

We believe that colored Braille is similar to colored graphemes in that Braille involves sixty-three different *symbols* derived from various combinations of the six dots in the Braille writing cell. For MD, each of these sixty-three symbols has its unique and consistent color. One estimate is that 50% of those who become blind in childhood develop colored hearing.[18] MD recently spoke before a group of "nearly a thousand visually impaired persons," mentioning her synesthesia for the first time. Six other blind persons from that group volunteered experiences similar to hers.

Touched by Sound

Carol Crane has sound → touch synesthesia, meaning that musical instruments evoke tactile impressions of stroking, pressure, warmth, and tingling. For example, she consistently feels the sound of different instruments on different parts of her body—guitars "brushing" her ankles on up to her shins, violins "breathing" on her face, cellos and organ "vibrating" near her navel. Trombones feel like a "dull throbbing" on the back of her neck, which is why she dislikes New Orleans jazz. It appears that an instrument's unique timbre is responsible for a distinct touch, and not all instruments induce feelings. "I love going to the symphony, but afterwards I am exhausted," she says, as if the experience were emotionally draining.

Although Carol's sensory combination is rare, it is not really odd: music often triggers physical sensations in listeners. Think of goose bumps, flushing, or tingling sensations caused by soaring or creepy music. Such cross-sensory stimulation in nonsynesthetes is one more reason to think that

Figure 2.10
"Telephone Ringing," one of Marcia Smilack's reflection photographs whose visual shapes evoke sound.

synesthetes may use, in an exaggerated way, neural circuits that are present in all of us.

Hearing with the Eyes

Marcia Smilack is one of those unusual synesthetes who hears what she sees. Most important to her is visual configuration, meaning a shape that possesses sound. She photographs reflections of sound–shapes, and we will say more about her later when discussing synesthetic art. Figure 2.10 illustrates a shape that evokes the sound of a ringing telephone.

Lidell Simpson also experiences this type of synesthesia. For him, blinking lights or moving objects cause him to hear a sound in his head. He cannot, he reports, look at the flashing lights on a radio tower or the trees moving by from a moving vehicle without actually *hearing* the visual experience.

Recently, Melissa Saenz of Caltech has been studying this visual movement → sound synesthesia. To her surprise, she discovered it to be more common than previously suspected: she was able to find four such synesthetes merely by asking the right questions around her colleagues and acquaintances. If this high prevalence proves true across larger samples, it means this type of synesthesia will soon be promoted out of the "unusual types" section of our book. In the meantime, Melissa has been proving the perceptual reality of this synesthesia by showing that visual movement→ sound synesthetes perform better than controls at visual tasks involving the judgment of sequences in time. The auditory system is much better than the visual system at these sorts of fine-timing tasks, and it appears that this form of synesthesia allows vision to take advantage of the auditory system's fine resolution.

An Endless Variety?

As is evident from viewing table 2.1, there are many rare combinations of synesthesia. Some of the more unusual ones include temperature → sound, musical notes → taste, touch → taste, and smell → sound.

Common Diagnostic Features

Despite the broad variety of synesthesia's manifestations, all types share certain general characteristics. Long before neuroimaging was available, Richard devised a clinical diagnosis of synesthesia based on five common features partly as a way to distinguish deliberate contrivances and artistic ideas of sensory fusion from the perceptual synesthesia that occurs automatically without an external cause (such as LSD) or a brain abnormality (such as temporal lobe epilepsy).

Automatic and Involuntary

Unlike imagery, which is willed, synesthesia happens to you. It is easy to show synesthesia is automatic by inducing Stroop interference, so named after the researcher J. R. Stroop, who first discovered it in 1935.[19] Basically, naming the ink color of colored words such as "BLUE" is difficult because the answer (orange) conflicts with the semantic meaning of the word "blue." Stroop interference shows that even unconscious exposure to something can directly affect reaction time.[20]

Take a person who sees 7 as yellow and 9 as blue, and then give her the task of saying a math solution out loud followed by naming a color square (see figure 2.11). Having to answer "7" and then "yellow" is congruent with her synesthesia, which unconsciously primes her to respond much

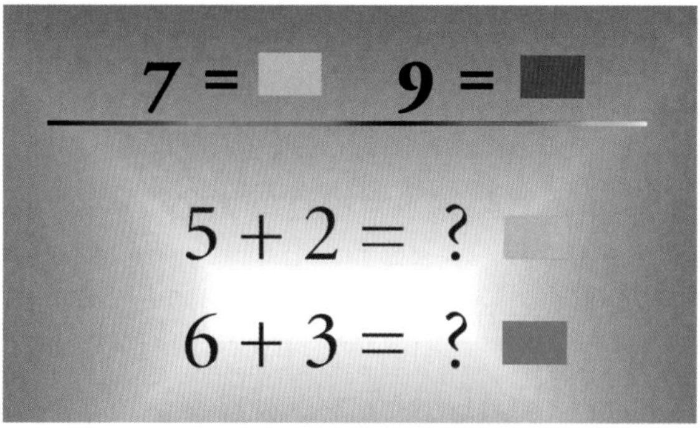

Figure 2.11
Color priming. Reaction times for answers that are congruent with the color of the automatic synesthetic photisms are faster than those for incongruent colors.

faster than controls. However, the automatic blueness of 9 interferes with naming the green square, tripping her up and slowing her down compared to controls.[21] Other approaches have used *cross-sensory* Stroop interference—for example, a "red" sound interfering with naming a green target patch.[22]

Stroop interference proves that synesthesia is automatic, but it does not by itself tell at what stage the interference takes place. Is it in early, unconscious processing stages, or instead during the later, consciously deliberate selection of a response? To understand more about the stage at which colors and graphemes interact, several research groups have sought tests in which synesthesia changes one's performance.

One of the first such tests asked whether synesthetes are faster in finding hidden patterns of letters or numbers than nonsynesthetes. Vilayanur Ramachandran and Edward Hubbard in San Diego asked the following question: Given that a red pattern against a green background will be immediately detected by a normal viewer (i.e., the shape will instantly be discriminated from the background without effortful searching), will a synesthete immediately be able to see, say, a pattern of 2s hidden within a field of 5s (figure 2.12)? Their expectation was that nonsynesthetes will have to search a while before finding the 2s, whereas synesthetes who see 2s as differently colored from 5s will quickly say, "Oh, there's a red triangle," or whatever the shape happens to be. For this to work, it would mean that synesthesia occurs very early in the chain of perception, so early as

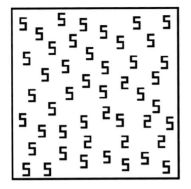 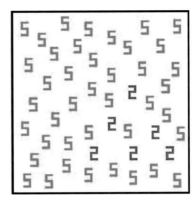

Figure 2.12

A field of 5s in which a pattern of 2s is hidden (*left*). Despite initial excitement about the possibility, synesthetic colors do not instantly pop out from the background (*right*) before the numbers are individually attended. However, synesthetic colors may allow more efficient anchoring of found numbers, allowing synesthetes to discern hidden patterns slightly faster than controls. (From Ramachandran and Hubbard [2001], with permission.)

to happen before subjects are even consciously aware of seeing a stimulus. As it turns out, despite initial excitement, synesthetes are generally *not* able to spot the hidden shape any better than nonsynesthetes.[23] This result is important, because it demonstrates the stage at which synesthetic colors are perceived: the number must be *attended to* for it to be synesthetically colored—it does not evoke color before the viewer is conscious of the number's identity.

Even though synesthetes do not have the magical power to instantly spot a hidden pattern, some evidence suggests they might be slightly faster than controls at this task. The explanation appears to be that once they perceive an oddball 2 in the matrix of 5s, the color helps them to anchor that finding as they go on searching for the next 2s (or alternatively, to more quickly reject distractors). This makes their searches more *efficient* compared to those of nonsynesthetes. Thomas Palmeri and colleagues[24] at Vanderbilt demonstrated this strategy by gradually increasing the number of elements in the matrix. Thus, rather than occurring preconsciously, synesthetic color appears to be bound to a visual form just as it is being recognized. This still means it occurs rapidly and very early in the perceptual chain—just not before awareness of the letter or number.

Another way to prove that synesthesia is perceptual is by using a physiological marker such as pupil diameter rather than a behavioral measure such as reaction time. For example, the pupil dilates more when viewing incongruently colored graphemes than congruently colored or black ones.[25] Pupil diameter is controlled by the autonomic nervous system; thus, measuring its size by infrared methods confers the advantage of not having to rely on verbal reports. The pupil provides a privileged window on subjective first-person experiences. Increasing diameter is thought to mark processing load and thus attention, and "wrong" pairings arouse attention more than congruent ones do.

Spatially Extended

Some synesthetes are adamant that what they experience is projected externally outside their bodies, sometimes on a "screen," whereas other synesthetes say that their perceptions are located in the "mind's eye." Careful questioning indicates that involuntary synesthetic experience differs from both seeing and imagination, which is willed. Whether synesthesia is frankly projected or not in a given person, what seems evident is that it has a sense of *spatial location*.

That is, synesthetes speak of "going to" or "looking at" a certain location to attend to whatever it is they experience. Sometimes, the spatial sense is literally outside the body—Michael Watson typically "reached out" at arm's length to palpate the shapes and textures he felt. When synesthesia is projected, it is experienced close to the body within limb's reach, never farther away (technically this is called "peripersonal space"). As we have seen, spatial extension is particularly evident in number forms, and we have already given examples of individuals who see music projected on a "screen" close to the face.

Jean Milogav likens her experience to a ticker tape:

You know on Times Square how they have that electric band with the news? That is exactly how it is in my head. Any word that comes in flows right through me in color. That's exactly how it is. It goes fast, of course, I mean I haven't got time in a conversation to think of everything—but it's simply there. If I want to I can stop at a certain word and look at it.

Somebody says to me, "How is your dog?" First I see the word DOG in color, then I think of my dog. That's how it goes. The color always comes first before I can think of the thing.

Consistent, Elementary, and Specific

What synesthetes experience remains generally consistent over time. Given the same stimulus, they nearly always experience exactly the same perception. This has been demonstrated repeatedly with test–retest situations—usually given without warning—for periods of many years. In fact, synesthesia researchers regard consistency over time as a "test of genuineness."[26] Also, given an array of choices on a matching task, synesthetes pick only a few, whereas controls distribute their choices over the available range.[27]

Synesthetic experiences are elementary: sensations have qualities like warm–cool, jagged–smooth, bright–dark, sparkling–steady rather than pictorial or highly elaborate. For example, synesthetes don't say, "This music makes me see sheep gamboling in a meadow." The generic perceptions of synesthesia constitute what are called "form constants," which may be possible building blocks of perception and which we discuss below.

Synesthetic perceptions are a part of the whole cloth of experience. As Deni Simon explains it:

The shapes are not distinct from hearing them—they are part of what hearing *is*. The vibraphone, the musical instrument, makes a round shape. Each note is like a little gold ball falling. That's what the sound *is*; it couldn't possibly be anything else.

Finally, synesthetic experiences are specific. Those who taste phonemes describe very precise tastes. Given the biggest box of crayons or a large array of paint chips, synesthetes are dissatisfied with the choices offered because none exactly matches the color they perceive. When using a computerized color selector, they will often spend many minutes finding the exact color match out of 16 million that are available. As Galton noted over a century ago:

Seers are invariably most minute in their descriptions of the precise tint and hue of the color. They are never satisfied, for instance, with saying "blue," but will take a great deal of trouble to express or match the particular blue they mean.[28]

For instance, synesthetes use a lot more words to describe colors than do nonsynesthetes—495 compared to fifty-eight according to a study by Julia

Table 2.4
Shades of green for nonsynesthetes and synesthetes

Nonsynesthetes (5 shades)		
green	dark green	lime green
emerald	Avocado	

Synesthetes (52 shades)		
green	jade green	pea green
murky green	khaki–ish pale green	pear green
grass green	leaf green	strong green
apple green	spring leaf green	green/dark
blackish dark green	light green	very dark green
sherwood green	lime green	very dark green/blue
bottle green	lime pale green	dark green almost black
bright green	live green	watery green
bright forest green	medium green/bluish	green with yellow
cos lettuce green	mid green	yellow dirty green
dark green (yellow/red)	mid/dark green	green/brown
dark blackish green	mossy green	greenish
dull light green	green and yellow	greenish bronze pulsating
fir tree green	forest green	greenish bronze strong
grey green	olive green	dark greenish brownish
grayish light green	olive/mustard green	muddy green
hard angular green	pale green	dark green
	pale transparent green	

(From Simner et al. [2005], with permission.)

Simner and colleagues.[29] Comparing variants in their descriptions of "green," for example, Julia counted fifty-four different shades by synesthetes compared to only five shades used by nonsynesthetes (see table 2.4). Rather than simply possessing a more voluble color vocabulary than nonsynesthetes, we believe synesthetes experience qualitatively more varied color experiences and merely attempt to describe them accurately. As such, the quantitative differences in synesthetes' reports are an additional indication of their genuineness.

Highly Memorable

When asked what good synesthesia does, a common response is, "It helps me remember." Perhaps because synesthetic percepts lack semantic meaning, they are easily and vividly remembered, often better than the triggering stimulus. "She had a green name—I forget, it was either Ethel or Vivian." In this example, the actual names are confused because both are green, but the synesthetic greenness is recalled.

The relationship between synesthesia and heightened memory is, of course, well depicted in Luria's *The Mind of a Mnemonist*. His subject's memory was "limitless and without distortion" largely because of the fivefold synesthesia that accompanied every event. Describing his process of recall, Luria's synesthete reported,

I recognize a word not only by the images it evokes but by a whole complex of feelings . . . it's not a matter of vision or hearing but some over-all sense I get. Usually I experience a word's taste and weight, and I don't have to make an effort to remember it—the word seems to recall itself. What I sense is something oily slipping through my hand . . . or I'm aware of a slight tickling in my left hand caused by a mass of tiny, lightweight points. When that happens I simply remember, without having to make the attempt . . .[30]

In our experience about 10% of synesthetes experience eidetic images, or what is popularly called "photographic memory."[31] The co-occurrence of synesthesia and eideticism was first noted by the German psychologist Jaensch[32] in 1930. Others have studied the relationship more recently.[33] An eidetic image recreates previously seen objects or events with great clarity, either immediately after they have been viewed or after a considerable period of time. Like some synesthetic percepts, eidetic images are spatially extended in that eidetikers behave as if they are scanning an externally projected image. Often they will look at some plain surface that serves as a convenient background.

Eidetisicm can be diagnosed as follows.[34] Using colored squares viewed against a neutral background, the examiner demonstrates afterimages such that the subject learns to distinguish an afterimage (which moves with eye movement) from an eidetic image (which does not). Unlike eidetic images, afterimages fade rapidly, require long viewing times to establish, and are negatively (complementarily) colored. Criteria for an eidetic image are that an image (1) must be reported, (2) must be positively colored, (3) is projected onto the easel rather than being located in the head, (4) is described in the present tense, and (5) is associated with eye movements appropriate to the location of objects in the scene.

Many reports on eideticism are in children, with an estimated 8% of American elementary schoolchildren having some decree of eidetic ability;

the prevalence estimate for strong eideticism in adults falls to 0.1%. As with synesthesia, we feel the drop-off may be due to the ongoing reorganization of the maturing juvenile brain.

Eidetikers easily perceive three-dimensional objects induced by random dot stereograms. This well-known tool for studying depth perception works as follows. Test squares consist of dots ranging from 10,000 to 1 million in number. One pattern of dots is shown to the left eye and a different pattern to the right eye. Binocular fusion then occurs in the brain, causing the subject to see a three-dimensional object floating above the test surface. Noneidetic individuals are unable to perceive the stereoscopic effect when the two different dot patterns are projected as little as 150 milliseconds apart. Eidetikers, on the other hand, can successfully fuse the two images even after a time lag of days![35]

The novelist Vladimir Nabokov was both synesthetic and eidetic.[36] His novels are notoriously autobiographical, so it is no surprise that he endowed several fictional characters with synesthesia. Among them is Van in *Ada*, who describes how color and number forms help him remember. "Synesthesia, to which I am inordinately prone, proves to be a great help in this type of task."[37] In *Speak, Memory* eidetic recall is a recurring subject, and recurring phrases in Nabokov's lexicon are, "I see myself," "I see," "I note," and "I distinguish." In passage after passage the past comes back to him in the same spatial–temporal perspectives as when he first encountered it. Chess games, for example, repeatedly show up in his novels with visuospatial specificity, and in *Pale Fire* he claims he can "order . . . photographs" to be taken, and can "reproduce" their exact image later on.[38]

Loaded with Affect

Synesthetes often gush over trivial tasks such as remembering a name or phone number, calling it "gorgeous" or "delightful," whereas mismatched perceptions—such as seeing a letter printed in the wrong color ink—can be like fingernails on a blackboard. Even when a percept is unpleasant or overwhelming, synesthetes love their experience and would hate to lose it.

The blind synesthete MD says,

I can do mental mathematical computations accurately *and with pleasure* . . . I find it *easy* and *satisfying* to picture street maps and am a good navigator. The maps that

I see are colorful and perhaps that's what makes map visualization both easy and *pleasurable*.

WW is a professor of neuropathology with colored phonemes and number forms:

Let me say that this is a *delightful* trait to have. I tend to use it consciously and unconsciously to help me remember correct sequences . . . neuropathological classifications, names and locations of anatomical structures (especially neuroanatomical structures—you should see the *beautiful array of colors* in the brain!).[39]

Some people have the "wow" factor more than others. Sean Day, who tastes milk as blue, is more amused than discomfited at the supermarket when a jug of 2% milk has a purple or green cap instead of a blue one. He is not wowed very much unless he takes the trouble (when his wife is away) to cook an inventive dish that is satisfyingly a favorite particular shade of blue. (Yet note that his preference for blue tastes is in itself emotional.) At the other extreme, Chris Fox's polymodal synesthesia is highly colored emotionally. For him, graphemes lead to color, smell, gender and personality, whereas sound and sight trigger taste, touch, shape and color:

I navigate through a rather incredible world. It's matter-of-fact in that it's always been that way, but it is not matter-of-fact in that new and strange synesthesiae always surprise me. The emotions in synesthesia are invariably strong and I take care to coordinate with them. Overload is a problem for me.

I have a full emotion for every color. I am careful to wear the colors of clothes that accord closely with my actual emotions at the time I put them on. If my mood changes during the day, I feel lopsided, out of synch between my own emotions and the emotions of the colors I am wearing or surrounded by.[40]

Jean Milogav reacts strongly to the color of names. Her niece was expecting a baby and wrote to say they would name it Paul if it was a boy. Jean became distraught because

the name Paul is such an ugly color, it's gray and ugly. I told her, "anything but Paul." And she couldn't understand why and I said, "It is such an ugly color, that name Paul." She thought I was out of my mind. At last I thought it really isn't my business, she can do what she likes. The name probably isn't that bad, but in my mind it's very awful. And that influences how I feel about people.

She hates her own name "Jean" and calls herself "Alexandra" among her family because, "The blue in the *A* is very nice," even though "I like blue not that much . . . The name Alexandra is such a pretty color and that's why I like it and that's why I always use it."

Another woman quit going to her parents' church because

I could not deal with the noxious colors and sounds of their music. Of course, I didn't really tell them why I was not going—the synesthesia part—it's just that I wasn't planning on torturing myself with dreadful looking music any longer.

Obviously, nonsynesthetes do not have such strong emotional reactions over such trifling matters. But for synesthetes, the emotional loading directly influences likes and dislikes. Sherelle Smith is a vocalist who composes some of her own tunes. She chooses key colors according to how she wants to feel while singing:

Romance is definitely purple (E or E-flat) . . . A or A-flat (two varying shades of blue) is for when I just want to sing something cool and mellow. The key of C is green, and though it's the best key to sing in it rubs me the wrong way. When I write in this key I usually end up modulating to another key at the end. Yet when I think of the serene green fields of England where I lived as a child I don't understand that at all. It seems I don't like "hearing" green as much as I like "seeing" green. But I definitely like seeing and hearing most shades of purple so I write a lot of songs in those keys.[41]

The Form Constants

In discussing synesthesia's spatial extension above we mentioned form constants in passing. These elementary patterns of spatial configuration were discovered starting in the 1920s by the German psychologist Heinrich Klüver. He induced them with mescaline to better understand the subjective experience of visual hallucinations. However, he quickly discovered that subjects were easily overcome by awe and the "indescribableness" of what they saw. The novelty of the visions and their vivid coloration captured subjects' attention more than the arrangement did, causing them to give way to cosmic *interpretations* rather than factual descriptions.

Once Klüver trained subjects to carefully introspect and hone it down to the sensory essentials, however, he identified four basic configurations he called tunnels and cones, central radiations, gratings and honeycombs, and spirals. These constitute the form constants (figures 2.13 and 2.14). Variations in color, brightness, symmetry, replication, rotation, and pulsation provide further gradation of the subjective experience. As the figures show, form constants are often symmetrical. This is why we said that when listening to music, synesthetes do not visualize something like a pastoral meadow with sheep gamboling through it—rather they perceive cross-hatchings, zigzags, circular blobs, cobwebs, and geometric shapes. It is the reason Michael Watson's tasted shapes were largely geometric in nature.

Klüver suggested that a limited number of perceptual frameworks may be inherent in the fabric of the central nervous system:

The analysis . . . has yielded a number of forms and form elements. . . . No matter how strong the inter- and intra-individual differences may be, the records are remarkably uniform as to the appearance of the above described forms and configurations. We may call them form-constants, implying that a certain number of them appear in almost all mescal visions and that many "atypical" visions are upon close examination nothing but variations of these form constants.[42]

Klüver's work was replicated and extended by others.[43] Unaware of Klüver's work, yet another researcher[44] rediscovered the repetitive elements of hallucinations and argued similarly that there are "certain constancies" the *visual system itself* contributes to illusory and hallucinatory phenomena as well as to everyday objective experience. The idea is that some basic anatomical or functional unit in the brain causes it to favor certain fundamental constructions of perception. Generic-looking images similar to the form constants also occur in a variety of nonsynesthetic settings—for example, during the aura phase of migraine, sensory deprivation,[45] the momentary twilight immediately preceding sleep (hypnagogic hallucinations), psychosis, and delirium.[46] Common conditions inducing the latter include fever, drug reactions, and low blood sugar.

Form constants—which are a product of the brain—differ from entopic phenomena, meaning "within the eye." That is, if you press on or rub your eyeballs the mechanical pressure stimulates the retina, causing you to see flashing lights or streaks, sometimes colored. Some people have an unusual ability to see their own retinal blood vessels, which produces a cobweb pattern. Small ghostlike circles and arcs called the muscae

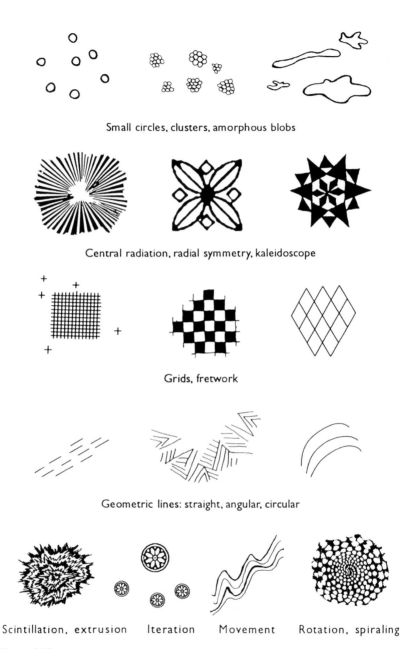

Small circles, clusters, amorphous blobs

Central radiation, radial symmetry, kaleidoscope

Grids, fretwork

Geometric lines: straight, angular, circular

Scintillation, extrusion   Iteration   Movement   Rotation, spiraling

Figure 2.13
Examples of Klüver's form constants.

Figure 2.14
Subjects' drawings of central radiation, spiral, and tunnel form constants. (From Cytowic [2002], with permission.)

volitantes are actual red blood cells coursing through vessels near the retina's macula. Glaucoma causes halos and rainbows around objects, whereas floaters and other opacities in the jelly-like vitreous humor behind the lens also cause blobs or cobwebs. Unlike the form constants, which are fixed in location relative to the environment, entopic images move with the eyes.

In the BBC documentary "Orange Sherbet Kisses," synesthetic artist Carol Steen and synesthetic art critic Bill Zimmer of *The New York Times* are shown viewing Kandinsky paintings. Kandinsky was a synesthetic painter who coupled four senses: color, hearing, touch, and smell. In the film Bill Zimmer not only points out form-constant shapes in Kandinsky's canvasses but also ones he experiences himself in response to sound. Figure 2.15, by a different synesthete, illustrates the shapes seen in response to environmental sounds—thunder, a bang, a clang, and a click.

The configurations of the form constants are not just visual but, more broadly speaking, *sensory* configurations manifesting in any sense capable of spatial extension. Figure 2.16 illustrates how touch and vestibular sensation can be localized in space just as vision is. Michael Watson expressed a spatial extension of touch—in describing the taste of mint as "cool glass columns," he spoke of "reaching through" rows of columns and of "tuning my hand to rub the back curvature." Even Michael's ordinary sense of taste had a spatial location, as he often remarked that he tasted flavors in different places in his mouth. A spatial extension of taste also appears in literature reports of colored taste.[47]

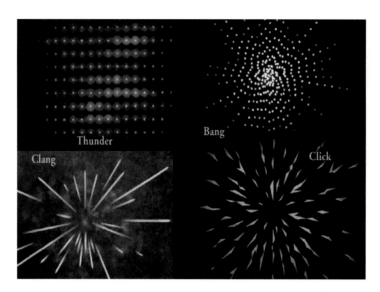

Figure 2.15
Subject's drawing of common environmental sounds, illustrating form constants of a grating ("thunder"), a combined spiral and central radiation ("bang"), and two central radiations ("clang" and "click").

If we asked, "How many readers are fond of smoke and explosions?" we suspect few would respond positively. But if we instead asked, "How many of you enjoy fireworks?" the affirmative response might be unanimous.

Why do we enjoy fireworks so much? Millions of pounds of entertaining explosives go up all over the world, and millions of people turn out to watch them. What are these colored lights, flashes, and bangs? They do not represent real things in nature or remind us of anything on an intellectual level. They are abstract yet provoke a strong emotional reaction, inducing millions to watch and walk away satisfied, exclaiming, "That was wonderful!" without being able to say exactly what "that" was. No other form of abstract visual expression is as popular.

It may be that the form constants can help explain the satisfying appeal of something as unnatural as fireworks. The connotation of the term "constant" gives a false impression that what is perceived is invariant and stationary, when in fact the elements making up a configuration are highly unstable, continually reorganizing themselves in an incessant interplay of concentric, rotational, pulsating, and oscillating movements by which one

Figure 2.16
Form constants apply to any sense that can be spatially extended. The drawing illustrates the fusion of visual–touch–vestibular hallucinations during a migraine aura wherein the subject's legs feel like they spiral, echoing the spiral rotation seen visually and felt vestibularly (i.e., the tilted window).

pattern replaces another. The kaleidoscopic transition occurs at approximately ten movements per second.[48] We know from the study of chaos theory (nonlinear dynamics) that self-organizing systems—such as the brain—are far from equilibrium, a status that underlies their capacity to change radically and unpredictably. In the brain, this circumstance might underlie the kaleidoscopic and scintillating transformations that synesthetic individuals report as part of their subjective experience.

3 Don't It Make My Brown I's Blue?

If you happen to tell a friend with number → color synesthesia that your telephone number is 713-555-8240, she may remember it as ■■■■ – ■■■■ – ■■■■■. (As one synesthete said about David's phone number, "It's not bad, but I wouldn't wear it.") Number → color synesthesia goes something like this: when a synesthete sees a **6** printed in black ink, she knows it is black and sees it as black, but she also has the experience of greenness. That experience of green is automatic and involuntary. For some, the experience is internal (green in the mind's eye); for others, the color may have a location (say, superimposed on the letter). Typically, it is a little disconcerting for a synesthete to view a letter in the "wrong" color—for example, looking at a red **6** when it seems to that individual that only **3** can be red.

Similar to colored numbers are colored letters: these forms of synesthesia typically occur together and collectively are referred to as grapheme → color synesthesia. The color associations for each grapheme differ for each synesthete and typically have very precise hues.

We earlier mentioned David Starr Jordan's 1917 report of grapheme → color synesthesia entitled "The Colors of Letters."[1] Jordan, a Stanford psychologist, presciently wrote, "It has been misunderstood by writers, who have imagined that the peculiar individuals having this trait actually see the color on the letter, which is not the fact. It is a mental association, not a false vision." This description holds true today, even though a minority of grapheme → color synesthetes do experience their color as having a particular location. Even in that case, synesthetic colors are readily distinguishable from real colors in the outside world. We return to the different ways of experiencing colors below.

Dr. Jordan, himself a synesthete, understood that synesthesia arose spontaneously, without teaching or outside suggestion. When he asked his eight-year-old son Eric, "What is the color of A?" Eric responded without hesitation that A was red. Jordan jotted down all of his son's color associations in 1912 and never again brought up the issue until 1917. Between ages eight and thirteen, 42% of Eric's grapheme colors had changed somewhat. In chapter 1 we cited this case as an example of how synesthetic associations can change during childhood. Nevertheless, when Dr. Jordan asked Eric again for his colors (see figure 3.1), he found enough correspondence in the boy's color choices to satisfy himself that his son's color correspondences had not simply been made up on the spot.

As surmised by Jordan, the reality of synesthesia can be shown by the fact that synesthetes show greater consistency than nonsynesthetes during repeat testing of their color choices.[2] However, although Eric Jordan's consistency was judged to be high, the problem remained that there was no

	Eric Jordan, 1913	Eric Jordan, 1917	David Starr Jordan
A	red	red bright	brown red
B	bluish	gray	green
C	white	white	yellowish white
D	bluish	gray	blue
E	pale green	yellow	red
F	red brown	brown	pale scarlet
G	pale brown	yellow	pale yellow
H	green	yellow	brown red
I	black	black	leaden black
J	dark blue	greenish	leaden
K	brown	brown	lead violet
L	pale green	green	green
M	red	brown	lead blue
N	pale greenish	light brown	brown red
O	light blue	black	white
P	yellow	yellow	lead color
Q	pale red	red brown	bluish white
R	dark green	dark red	bright green
S	silvery gold	silver	bright yellow
T	white	silver	green
U	yellow	yellow brown	yellowish
V	silver	white	violet blue
W	red brown	brown	lead blue
X	silver	silver	scarlet
Y	silver	white	blue
Z	reddish	dark brown	scarlet

Figure 3.1

Letter-to-color associations for David Starr Jordan's son Eric in 1912 and again in 1917. In the final column, Dr. Jordan lists his own synesthetic colors. Note that the colors are fairly consistent across time for a given synesthete and are idiosyncratic between synesthetes.

Figure 3.2
A screenshot from the online software at http://synesthete.org. A letter or number appears on the right (a T in this case), and participants use the color selectors on the left to find the best match to their synesthetic color.

good way of quantifying his responses. Was "white" in one test the same as "silver" in later testing, or should it be counted as different? And if so, how much different?

The introduction of computerized color matching has allowed a precise way to address these issues. In one method, internal consistency of color choices is tested not across years but within a single session. In a test developed by David's lab, participants sit in front of a computer screen and are presented with a random letter or digit (see figure 3.2, and take the tests yourself at www.synesthete.org). Navigating the mouse over a color palette lets them choose one of 16,000,000 different colors that most closely match their synesthetic experience for a given character. After selecting a color, they are presented with the next letter or digit. A participant is presented with a total of 108 trials (the full set of graphemes A–Z and 0–9) three times in randomized order. The data are then analyzed for consistency: did the participant chose a similar color the first time she saw the letter T as well as when she saw it some time later? The distance in color space can be computed for an exact score of consistency (see figure 3.3). Through this testing method, synesthetes can be readily distinguished from control subjects who are asked to use free association and memory in choosing their colors.[3] Tests like these have been another tool in confirming synesthesia as a real perceptual phenomenon.

A number of mysteries remain about grapheme → color synesthesia. For instance, why do some synesthetes experience color for both letters and

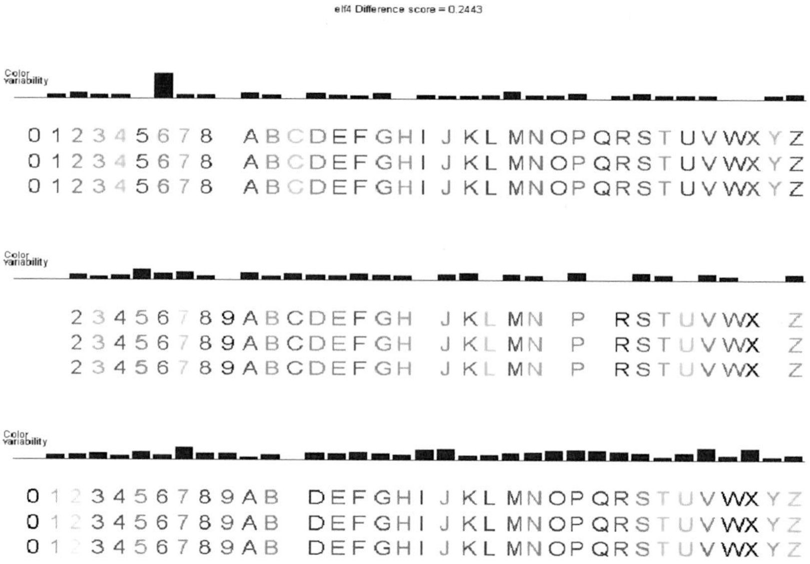

eif4 Difference score = 0.2443

Color
variability

0 1 2 3 4 5 6 7 8   A B C D E F G H I J K L M N O P Q R S T U V WX Y Z
0 1 2 3 4 5 6 7 8   A B C D E F G H I J K L M N O P Q R S T U V WX Y Z
0 1 2 3 4 5 6 7 8   A B C D E F G H I J K L M N O P Q R S T U V WX Y Z

Color
variability

2 3 4 5 6 7 8 9 A B C D E F G H   J K L M N   P   R S T U V WX   Z
2 3 4 5 6 7 8 9 A B C D E F G H   J K L M N   P   R S T U V WX   Z
2 3 4 5 6 7 8 9 A B C D E F G H   J K L M N   P   R S T U V WX   Z

Color
variability

0 1 2 3 4 5 6 7 8 9 A B   D E F G H I J K L M N O P Q R S T U V WX Y Z
0 1 2 3 4 5 6 7 8 9 A B   D E F G H I J K L M N O P Q R S T U V WX Y Z
0 1 2 3 4 5 6 7 8 9 A B   D E F G H I J K L M N O P Q R S T U V WX Y Z

Figure 3.3
Example numerals and alphabets from three grapheme → color synesthetes. Each
participant chooses colors three times for every letter and digit, presented in random
order. The color difference between successive choices is computed to produce a
consistency score.

numbers, whereas others experience it for letters only or numbers only?
Why are some graphemes colorless for given synesthetes? (For example,
the second synesthete in figure 3.3 has no color associations for 0, 1, I, O,
Q, and Y, whereas the third synesthete has no colors for C.) Indeed, synesthetes say that some colors are strong and vivid whereas others appear
"pale" or "washed out" as if the dial on a color television had been turned
way down.

These differences, while not totally understood, introduce a useful metaphor that we explore in chapter 9, namely, that the landscape of associations in the brain is mountainous, with some peaks poking above the cloud
cover of consciousness while others lie below. This metaphoric image raises
the possibility that the surprising cross-connections of synesthesia are
present in all brains—just at a lower elevation (and thus below consciousness). We explore this idea in depth in chapter 9.

One way to make progress in understanding synesthesia is to ask whether
the shape of this mountainous landscape can change during adulthood.

The answer appears to be yes, as we know from looking at synesthesia for letters from foreign alphabets. That is, what happens when a synesthete looks at foreign letters written in non-Roman scripts such as Hebrew, Arabic, Cyrillic, or Chinese? Typically, the newly learned shapes are not accompanied by a color experience. As Cassidy C writes, "they look like little black shapes on a white background. They have no color effect whatsoever." But most synesthetes report that once they have *learned* a language (even as an adult), the new graphemes take on colors. Often there is some correlation between the sound of a letter and its associated color in both alphabets. Together with the observation that synesthesia sometimes changes during adolescence, this implies that the peaks of the landscape can continue to shift through adulthood, which means that the brain retains the ability to change.

Combinations and Context

Figure 3.3 displays graphemes in their individual colors—but when letters become grouped into a word, the colors can change dramatically due to context. For many individuals, a word's first letter dominates the rest of the sequence, whereas others discern a blending in which all the letters influence one another. Some report that vowels tend to fade into the background under the dominating influence of neighboring consonants; in other cases, vowels inherit the shade of nearby colors. Some letters have more influence than others when appearing at the beginning of a word. For example, Cassidy C reports that the beginning letter I "can give an entire word a luminescent quality, and the consonants often lose some of their power in its presence." He also reports that the repetition of a letter in a word—as in the three Ss of synesthesia—influences the word's overall hue. In this case, the green Ss dominate and shift the rest of the letters towards green (see figure 3.4).

Typically the color of a word has less to do with pronunciation than spelling—so, for example, EROS and ARROWS would be quite different colors despite sounding similar. On the other hand, both the spelling *and* pronunciation matter for some synesthetes. Consider homographs, words with identical spellings but different pronunciations, as in "my smile is my best *attribute*" or "I *attribute* my success to my smile." For Krissa K, word shadings change according to where the stress is in a homograph (see table 3.1).

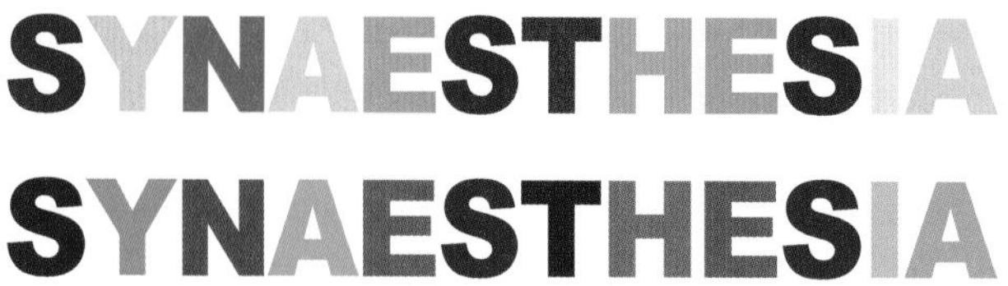

SYNAESTHES A
SYNAESTHES A

Figure 3.4

The repetition of a letter, as in the Ss in this word, can influence the color of other letters in the word. The word on top is drawn in the colors of the individual letters of Cassidy C, while the word on the bottom is how he perceives the word when he views it as a whole. Note that he has used the British spelling in this example, while we use the American spelling throughout this book.

Sensitivity to pronunciation seems to hold for only about 25% of grapheme → color synesthetes polled[4] and implicates an auditory component to their synesthesia. In other words, some grapheme → color synesthetes have an interaction with *auditory* parts of the brain, whereas for most their synesthesia is based only on the *written* letters, a difference we return to in chapter 9.

In many cases word *meaning* also influences color. For example, it is common for the word APPLE to be reddish, BANANA yellow, and ORANGE orange. Sometimes synesthetes find that colors change subsequent to their learning a new word definition. When Cassidy C encountered the new word "phthalocyanine," his synesthetic color experience was determined by the colors of the word's individual letters (figure 3.5, *top*). However, upon learning that phthalocyanine is the name of a vivid blue–green pigment, he now experiences the word as shown in the bottom half of figure 3.5.

Color associations usually provide a good method for remembering spelling, but they can occasionally lead to confusion in the case of similarly colored names (see figure 3.6). In Cassidy C's case, Mike and Dave have similar colors, as do Dan and Rob, leading to social uncertainty at cocktail parties.

As we see by the descriptions above, the rules of letter-to-color associations are varied: some synesthetes find that their colors are influenced by the first letter, others by vowels or consonants, some by pronunciation, and still others by meaning or context. This variety of experience is not yet fully understood, but in chapter 9 we will explore how natural variability in individual brains may underlie this variability of expression.

Table 3.1
Examples of homographs and their synesthetic colors

Attribute	My smile is my best **attribute**.	bright yellow
	I **attribute** my success to my perseverance.	darker, muckier looking, a kind of yellow-gray
buffet	Steamed clams are at the left edge of the **buffet** table.	reddish brown
	It takes a catastrophe every now and then to **buffet** the nation out of its laziness and complacency.	darker, shinier brown
compound	I earn **compound** interest on my savings.	light gray
	I'm worried that the weather will **compound** my troubles.	stays the same
desert	To **desert** the military is a crime.	muddy brown
	The Gobi is a large **desert** in Asia.	golden brown, like toast
present	All need to be **present** for a unanimous vote.	White
	He will **present** his ideas to the Board of Directors tomorrow.	stays the same
record	She played a vinyl **record** on her turntable.	dark blue
	Did he **record** the concert with his camcorder?	slightly lighter shade of blue
resume	**Resume** breathing or you will surely faint!	dark blue
	My **resume** highlights my extensive work experience.	lighter blue
voyage	I'm getting ready to take a **voyage**.	blue–gray
	Bon **voyage**!	stays the same
wind	How did we **wind** up in Kansas?	light gray
	The **wind** blew us here.	dark gray

Note. For a minority of synesthetes, different pronunciations of a single spelling can modify the associated synesthetic color, implicating an auditory component to their synesthesia. The right column shows the colors reported by Krissa K.

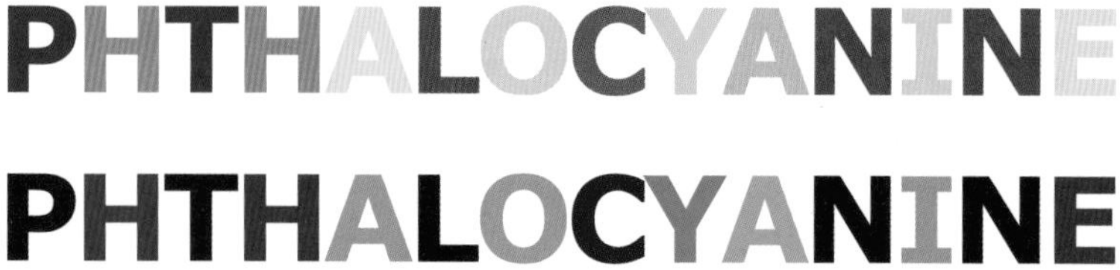

PHTHALOCYANINE
PHTHALOCYANINE

Figure 3.5
Cassidy C's synesthetic experience of a word before (*top*) and after (*bottom*) learning that "phthalocyanine" refers to a vivid blue–green pigment.

MIKE   DAN
DAVE   ROB

Figure 3.6
Similarly colored names in the case of Mike and Dave, and Dan and Rob, can be confused (Cassidy C).

Weekdays and Months

Colors for weekdays and month names, such as Thursday and February, tend to be determined by concept rather than spelling. In fact, it is more common to experience days of the week as colored than it is to simply experience colored graphemes. A weekday → color synesthete may tell you that Wednesday just feels magenta—but not because the letters W–e–d–n–e–s–d–a–y are magenta. A synesthete thinking about Tuesday may have an orange experience, perhaps, whereas Friday feels like a nice deep red (see figure 3.7).

What happens if someone has both letter → color and month → color synesthesia? Is a month's color determined by the spelling or the concept? The answer is both, depending on what you pay attention to. For example, Gizelle T, a native of Puerto Rico, learned English and Spanish simultaneously as a child. For her, color is often associated with the concept of the day rather than the spelling (e.g., January and its Spanish equivalent, *Enero*, are both the same color if she thinks about the winter month. But she was surprised to realize that when she attends to the spelling, her colors are sometimes different—for instance, when F–r–i–d–a–y is spelled, it is red, whereas *V–i–e–r–n–e–s* is blue.

Sunday	January
Monday	February
Tuesday	March
Wednesday	April
Thursday	May
Friday	June
Saturday	July
	August
	September
	October
	November
	December

Figure 3.7
One synesthete's weekday and month colors. For this subject the month of November was close to white; it is therefore displayed against a black background for easy reading.

Where Is the Color? Localizers and Nonlocalizers

We now return to an issue touched on earlier. Grapheme → color synesthetes experience a color when they see a letter—but *where* do they see the color? Does it have imagined spatial coordinates, hover in real space like a hallucination, or exist as a placeless inner experience of color? This is an area ripe for confusion, so we suggest some clarifying terminology.

In general, there appear to be different ways of experiencing colored letters. In many cases, the color is experienced as an internal sensation, or "feeling of color," the way you might visualize indigo if asked to strongly do so right now. In other cases, the color has a definite location either internally or externally, and here the experience is called a photism. (Some individuals with sound → color or taste → color synesthesia report localized photisms that obscure real, physical space the way hallucinations do, or else they say things look tinted by a transparent "overlay," as if looking through a piece of colored cellophane. Here, however, we are talking only about colored graphemes and have yet to encounter grapheme → color synesthetes who perceive colors opaquely.)

To understand the difference between a localized photism and a hallucination, imagine a shiny red apple on the table in front of you, just within arm's reach. Now, if we were to ask you if the apple was "in your mind's eye" or "out in the world," what would you say? The answer should be that you see it in your mind's eye, since it is an imagined apple rather than a real one. On the other hand, especially if you are a talented visualizer, it also seems to be out there—after all, you envisage it in a specific location in space, and you may almost feel as though you can touch the spot. The apple is not a hallucination, but it does have a perceived location in real-space coordinates.

Understanding this difference is important, because in previous years some investigators sought to distinguish different types of synesthetes by asking whether their colors were perceived "in the mind's eye" or "out in the world." The problem is that for those who localize colors the phrasing is unclear, as in the apple question. When asked to indicate a location, some synesthetes answered "out there in the world," which led to the terminology by some researchers of "projector" and "associator" synesthetes.[5] This distinction was useful in spirit, but caused confusion in the popular press. Problematically, the term "projector" conjures up images of a movie projector and suggests a hallucination, while "associator" sounds to some ears like a memorized association, which synesthesia is not. In place of these terms, we suggest "localizer" and "nonlocalizer."

Although we use these shorthand terms occasionally, in our formal notation the distinction between these two types of synesthetes is simply this:

[grapheme → color] = a nonlocalizer who experiences colors with no specific location (the way your conceptions of "justice" or "friendship" have no particular location).

[grapheme → color, location] = a localizer who experiences synesthetic colors as belonging to a specific location.

Note that both types can tell the difference between synesthetic colors and real colors (that is, they do not make the mistake of insisting that a black 3 is actually colored). And they can tell the difference between synesthetic colors and objects in the real world. In other words, synesthetic experience—even for localizers—is neither film projection nor hallucination.

Does it seem surprising that an imagined location would not interfere with real objects? To see why this is not strange, put your imagined red apple in the same location as your real coffee cup. You'll see there is nothing impossible, or even particularly confusing about two objects, one real and one imagined, sharing the same coordinates. Hallucinations, in contrast, can block out or override perceptions in the real world.

Distinguishing between localizers and nonlocalizers is important for understanding the neural basis of synesthesia (chapter 9). A popular account of grapheme → color synesthesia is based on the observation that a region of the brain responsible for color perception (area V4) lies close to the region involved in grapheme recognition. One suggestion is that synesthesia stems from a cross-activation of the adjacent areas due to either overconnectivity of neural areas[6] or a lack of inhibition. Consistent with this idea, Julia Nunn's team in Britain used functional magnetic resonance imaging (fMRI) to show that synesthetic color induced by spoken words produced brain activity in brain areas normally associated with color perception.[7]

For localizers, then, a three-way interconnection is implicated between brain areas that code for (1) graphemes, (2) colors, and (3) spatial areas. For nonlocalizers, only a connection between the first two areas is implicated. The details of these areas will be discussed in chapter 9.

Do We Need Attention to Trigger a Color?

An open question is whether a grapheme needs to be *attended to* in order to evoke synesthetic color. The question is harder to answer than it appears. Saying that synesthesia is automatic and involuntary suggests it just passively "happens" without the need to pay any attention at all to stimuli. When reading, for example, synesthetes attend to word meaning and phrases just like the rest of us do rather than to each individual character. Yet Carol Steen and Megan Timberlake speak of seeing "Technicolor on the page" even though they see the printing in black ink, whereas Jean Milogav likens her color experience to a ticker tape speeding through her head. "When I read or listen to conversation," she says, "color just flows through me." If she wants to she can "stop at a certain word and look at it," examining the colors in detail.

These individuals' first-person experience would seem to confirm that synesthesia is always "just there." However, it may be that different *levels* of attention can be devoted to it. This can account for the "blur" or "flow through" of color that the synesthetes in our example speak of. Other synesthetes concur, claiming similar experiences. One such individual, Julie Norich, was featured in the BBC documentary, "Derek Tastes of Earwax," watching a ticker tape of color flow through her as the announcer rattled off railway departures. When such individuals stop to focus on individual characters, colors then become distinct and definite. In other words, a synesthete's purple J isn't strongly purple all the time: it is most purple when its J-ness is being attended to. The distinction is similar to that between looking and seeing. Therefore, it is difficult for either synesthetes or researchers to know whether a grapheme has much of a color when it is not attended to. Note that this observation in no way changes the description of synesthesia as involuntary.

If this concept of attending to a synesthetic color seems strange, ask yourself what is the exact color of the coffee mug in front of you. This is easy to answer, but you may not have fully appreciated the answer before you considered the question and really paid attention. Try looking around your present surroundings and ask yourself if you were aware of the exact color of each object before you attended to it. In the same way that a red Coca-Cola can in front of you will always appear red when you attend to it, you can attend more or less strongly to that aspect of it. At times you will pay more interest to its content or temperature, but anytime you ask yourself about the color, the answer is red.

This issue is important because it influences the direction of synesthesia research. In earlier days, before this point was fully appreciated, it was thought that synesthetes should be able to more readily pick out hidden patterns that nonsynesthetes could not discern (refer to the 2s hidden among 5s in the last chapter). The dashing of this expectation based on later experiments was an important lesson in thinking about synesthetic experience. The experiment fails to work on every synesthete because the 2s in the field are not colored until they are recognized and attended to. The important point is that for the overwhelming majority of synesthetes it is not the *shape* the letter casts on the retina that generates color but instead the consciously attended concept of the letter. The letter must be perceived consciously before a color is experienced.

Concept Rather than Shape

It now seems clear that for the majority of synesthetes it is the concept inherent in a grapheme that induces color—not the visual shape itself. To demonstrate this, note that capitalization and font style generally do not change an induced color: j, J, and *J* all evoke the same synesthetic color.

In fact, one can show that the concept determines the color by triggering different colors using a single shape. In a demonstration by Ramachandran and Hubbard, synesthetes reported when reading THE CAT in figure 3.8 that the synesthetic colors for the H and the A were different, even though the shapes of the graphemes were cleverly engineered to be identical.[8] Again, this shows that the colors are yoked to meaning rather than shape.

Similarly, in a stimulus known as a Navon figure, one can mentally switch between the forest and the trees, attending to either the 2 shape or the cluster of 5s (see figure 3.9). Synesthetes report experiencing the appropriate color based on what they attend to.

Whereas above observations depend on subjective reporting, a recent study showed even more objectively that color experience depends on interpretation.[9] That study used ambiguous forms like the one in figure 3.8 so that, for example, a 2 presented in a sequence of letters would be interpreted as a Z, whereas in a sequence of digits it would be perceived as a 2.

TAE CAT

Figure 3.8
In this example of letter ambiguity, the middle letter is graphically identical in the two words, yet most subjects will automatically interpret the first as an H and the second as an A. Synesthetes will experience the color according to the context.

Attending global   Attending local

Figure 3.9
Synesthetes report that attending to the global shape of the Navon figure (in this case, a 2 made up of small 5s) induces their color for the 2 (say, yellow), while attending to the local elements induces their color for 5 (red).

Researchers presented the ambiguous grapheme in different colors and measured how quickly synesthetes could correctly name the color. Exploiting Stroop interference, they set up the experiment so that the color was congruent when the grapheme was interpreted as, say, a Z, but was incongruent if the grapheme was interpreted as a 2. By looking at the speed of responses, they could verify that identical shapes triggered *different* synesthetic colors depending on how the shape was interpreted. Again, this shows that for most synesthetes it is not simply the visual shape but instead the meaning of the grapheme that determines their synesthetic coloration.

4 + 4 Is Enough

The next natural question is this: if the *interpretation* of a shape determines its synesthetic color, does the shape need to be seen at all? This question was tackled in a clever experiment by Mike Dixon and colleagues in Toronto.[10] Recall the synesthete from figure 2.11 who sees 7s as yellow and 9s as blue. Her task was to watch in sequence a number, then a plus sign, then another number. She then reported the mathematical result and named the color of a square presented to her. Having to answer "7" then "yellow" is easy and quick because for her 7 is synesthetically yellow. But having to answer "9" then "green" is difficult because 9's automatic blueness interferes with naming the green square and slows her down compared to control subjects. In other words, she was significantly faster at naming the color when it was synesthetically consistent with the answer to the problem.[11]

"Higher" versus "Lower" Synesthesia

When discussing how the concept rather than the shape triggers synesthetic color, you may have noticed us saying "for most synesthetes." For a small minority, the triggered colors are sensitive to the details of what is presented on the page. Hubbard, Manohar, and Ramachandran have suggested calling these "lower" synesthetes (indicating the low level of their association), as opposed to the more common "higher" synesthetes like those described in this chapter.[12]

For instance, they have found one individual whose synesthetic colors fade away when letters are presented at low contrast. For example, a black letter on a white background may trigger a blue color, as will a white letter on a black background—but a gray letter on a light gray background does not induce any color. Hubbard and Ramachandran found that as contrast was lowered, the connecting parts of the letters appeared to lose their colors first, an observation not yet understood. In samples from www.synesthete.org, we have found a loss of color with reducing contrast to be very rare, although one example is shown in figure 3.10.

The finding that synesthetic colors can change with contrast suggests that in a small minority of individuals the physical form of the grapheme on the page matters, not merely the concept. This again underscores how heterogeneous the synesthetic population is. Contrast dependence is not a general feature of grapheme → color synesthesia, but it is for some. As with the localizers and nonlocalizers discussed earlier, lower and higher synesthetes presumably have different details of neural crosstalk, a topic we explore in chapter 9.

Why the Letter A Tends To Be Red

Even though each synesthete's color associations are unique, careful analysis has uncovered some trends. For example, Simon Baron-Cohen and colleagues noticed that 73% of English-speaking synesthetes in their sample experienced the letter O as white.[13] In analyzing hundreds of self-reported colored alphabets, Sean Day found that A often turned out to be red, B

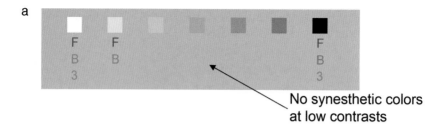

Figure 3.10

(*a*) One synesthete out of many hundreds reported that her synesthetic colors disappeared when letters were presented at low contrast. In this example, she was repeatedly presented with F, B, and 3 at different contrasts as indicated by the squares about the letters. At low contrasts, she reported no synesthetic color. (*b*) For the overwhelming majority of synesthetes, the same synesthetic colors are reported regardless of the contrast of the presentation.

blue, and C yellow.[14] Might there be common biases by which particular graphemes become associated with particular colors?

Delving into the question, Julia Simner and colleagues in Edinburgh found that more often than expected by chance, particular graphemes associate with particular colors.[15] They found that letter–color pairings tend to follow a trend. To understand that trend, let us first step back a few decades.

In 1969 linguists Brent Berlin and Paul Kay showed that human languages introduce color terms in a fixed order.[16] For example, a language discerning only two colors will use the terms "black" and "white" to distinguish light objects from dark ones. A language discerning three colors will employ the terms "black," "white," and "red." The next most likely term is "green" or "yellow," followed by "blue" or "brown." The next term to be added is equally likely to be "orange," "purple," "gray," or "pink." It seems synesthetes are sensitive to this typology because they tend to pair more common letters with colors that are introduced early. For example I,

O, and A are common letters and tend to pair up with black, white, and red, respectively.

When Simner's team ranked the frequency with which different letters appear in English (e, t, a, o, i, n, s, r, h, l, d, c, u, m, f, p, g, w, y, b, v, k, x, j, q, z) they found that more commonly occurring letters mapped to earlier color terms, and less common letters to later appearing color terms. For example, synesthetes tend to pair common letters (e.g., A) with common colors (e.g., red), and uncommon letters (e.g., Q) with uncommon colors (e.g., ochre). The reason early letters such as A, B, C . . . do not orderly map onto color terms is that as children we do not learn the alphabet sequentially. Rather, we learn the most commonly used letters first. Hence we find synesthetic pairing occurs based on a letter's frequency.

By contrast, we *do* learn numbers sequentially. In 1982 the Israeli psychologist Benny Shanon showed that the Berlin and Kay color typology roughly predicted synesthetic number–color pairings in that earlier learned low numbers were linked to early color terms whereas higher numbers, learned later, were linked with later color terms.[17]

Beyond color typologies, synesthetic associations are also influenced by a linguistic effect of initial letter priming that makes B likely to be blue, Y yellow, G green, and so on. Similarly, a handful of synesthetic pairings have been influenced by childhood objects such as refrigerator magnets[18] or coloring books.[19] The degree to which such imprinting influences color associations is currently unknown, but the fact that it occurs in some synesthetic alphabets provides an important clue about the development of the synesthetic brain. After all, as children, most of us played with magnets and coloring books, yet only a fraction of us synesthetically imprinted on such colors. For the overwhelming majority of the general population, colors were learned as accidental properties of the letters, and no associations stuck.

It happens that nonsynesthetes also follow rules when forced to make letter–color matches. They do not exhibit the relationship between letter frequency and color frequency but they do exhibit more predictable color–letter correspondences than is expected by chance.[20] This suggests that while grapheme → color synesthesia is experienced by a minority of the population, it may derive at least in part from mechanisms common to all brains. For example, nonsynesthetes tend to show initial letter priming, matching up B with blue, P with purple, Y with yellow, and so forth just above chance level. They tend to make these kinds of matches slightly more than synesthetes do.

Initial letter priming also holds for German-speaking nonsynesthetes. When a word shares an initial letter in both English ("white") and German ("*weiss*"), speakers of both languages tend to call W white. In cases where German and English write their color terms differently (e.g., English: "purple"; German: "*lila*"), the color tends to be triggered by a P in English and an L in German. This verifies a role of learned vocabulary in nonsynesthetes.

Of course, none of this means that every synesthete does exactly this all the time. It simply means that synesthetes are more likely than not to follow these trends. While none of the studies are able to predict any individual's particular associations, they provide clues to his or her under-lying patterns of association. Any neural theory of synesthesia will have to explain how high-frequency letters and high-frequency colors become bound together in the neural landscape.

Color as Memory Aid

Some tiny fraction of the population, called "mnemonists," have extraor-dinary, almost untaxable memories. One of the most famous of these was A. R. Luria's patient Solomon Shereshevsky—a fivefold synesthete. He attributed his prodigious ability to the extra sensations that accompanied every percept. It is not difficult to imagine how tagging letters or numbers with extra dimensions such as color and location would aid in recalling them later on.

The idea that colored numbers actually aid memory seems plausible—but how do we know if it is really true? In a clever study by Dan Smilek and colleagues in Canada, a synesthete called C was asked to memorize a matrix of numbers.[21] The numbers were presented in three different ways: in black ink, in colors congruent with C's number–color couplings, or in incongruent colors (see figure 3.11). C was asked to look at a matrix for one minute and then write down all the numbers she could remember.

The results? C outperformed a group of her nonsynesthetic peers when memorizing the black and colored matrices. But when presented with the

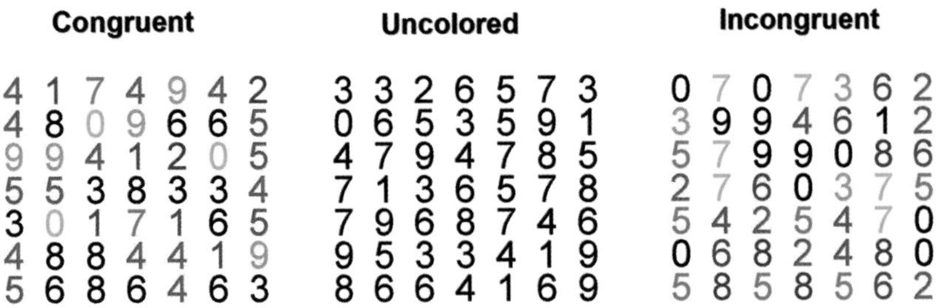

Congruent   **Uncolored**   **Incongruent**

```
4 1 7 4 9 4 2     3 3 2 6 5 7 3     0 7 0 7 3 6 2
4 8 0 9 6 6 5     0 6 5 3 5 9 1     3 9 9 4 6 1 2
9 9 4 1 2 0 5     4 7 9 4 7 8 5     5 7 9 9 0 8 6
5 5 3 8 3 3 4     7 1 3 6 5 7 8     2 7 6 0 3 7 5
3 0 1 7 1 6 5     7 9 6 8 7 4 6     5 4 2 5 4 7 0
4 8 8 4 4 1 9     9 5 3 3 4 1 9     0 6 8 2 4 8 0
5 6 8 6 4 6 3     8 6 6 4 1 6 9     5 8 5 8 5 6 2
```

Figure 3.11
Matrices of numbers presented to synesthetes (based on Smilek et al. [2002]). The numbers are displayed in congruent, incongruent, uncolored, or low-contrast arrays.

incongruent matrix, her performance hit rock bottom. After the test she asked the experimenters, "What have you done with my colors?" We also observe that the synesthete whose colors faded away at low contrast similarly performs poorly when trying to memorize a low-contrast matrix. In other words, assigning colors to numbers does in fact help you remember them.

When asked what good synesthesia does, a common response is, "It helps you remember." Perhaps because they lack semantic meaning, synesthetic percepts are easily and vividly remembered, often better than the triggering stimulus. While memorizing matrices is a special case, many people find that grapheme → color synesthesia helps them retrieve names. As mentioned in the last chapter, people often remember that someone possessed a name of a particular color. For example, when Linda DiRaimondo encounters an acquaintance she hasn't seen in a while, she says to herself, "That's a yellow name. Yellows are Ts and Ys." She does not know anyone whose name starts with Y, so by a process of elimination she goes through her mental list of names beginning with T until she remembers who her acquaintance is.

You've Got Personality

An advantage of the notation we introduced above is that many attributes can be associated with a grapheme, so the list can grow: [grapheme → color, location, personality, size, gender, shape . . .] and so on. And indeed, learned sequences such as letters and digits often trigger additional synesthetic experience besides color.

One relatively common form of synesthesia is the personification and genderfication of letters and numerals.[22] That is, graphemes take on personality, gender, and other qualities. For example, one synesthete quoted by Sean Day described the number 2 as "a shy, wimpy boy" and 9 as "a vain, elitist girl." She noted she dislikes certain number combinations such as 94 that "result in putting 4 (a plain but decent, hard-working older woman) and 9 together, as they greatly dislike each other and do not get along at all."[23] In a recent study by Noam Sagiv and colleagues,[24] one synesthete described some numerals:

2: He is dull as a rock and about as interesting. Not actively bad the way 6 is, just a fat ugly blah person. Tries to be teacher's pet also.
3: Kind of athletic and sporty, not super bright but . . . just not defined by intellect.
6: Big bullying type that sits in the back of class. Obnoxious, arrogant, and kind of stupid. Very politician-like and superficial.
7: Quirkily intelligent and really fun—wacky and creative but not annoying. Special, dynamic, like a glittering exciting person to know.

We have recently come to realize that Pythagoras (500 B.C.E.), the philosopher and mathematician, had number → personality synesthesia. To the best of our knowledge this has not been diagnosed historically, although it is immediately clear from his descriptions of numbers. The historian R. S. Brumbaugh notes that for Pythagoras:

Each number had its own personality—masculine or feminine, perfect or incomplete, beautiful or ugly. This feeling modern mathematics has deliberately eliminated, but we still find overtones of it in fiction and poetry. Ten was the very best number. . . .[25]

The historians who describe Pythagoras's peculiarities with numbers have apparently not been aware of the phenomenon of synesthesia, so his number personification is traditionally discussed as an interesting oddity. More commonly, those who cite Pythagoras's personified numerals are numerologists who would like to believe that Pythagoras had a cosmic insight into the true nature of numbers. A search on the Internet reveals hundreds of sites that take off from Pythagoras's number–personalities to construct New Age models of the nature of the universe. It is amusing that someone with a not-uncommon form of synesthesia could have such lasting effects on the details of what people subscribe to millennia later.

Although we now know that grapheme personification has existed for tens of centuries, this unusual phenomenon went unanalyzed in the English language until an 1895 paper written by a pioneering woman named Mary Whiton Calkins.[26,27] In the late 1800s, the young Calkins fought for access to seminars and classes at Harvard at a time when it was unheard of for a woman to earn a doctoral degree. Because of her gender she was not granted her doctorate, but nonetheless she went on to establish one of the first psychology labs in the United States at Wellesley College in 1891. She grew into a pioneer in the field, becoming the first female president of both the American Psychological Association and the American Philosophical Association.

By 1895, she had become interested in the issue of grapheme personification, which she referred to as "dramatization." Of the phenomenon she wrote that letters, numbers, and musical notes "often become actors in entire little dramas among themselves."[28] She tested 145 synesthetes and found that roughly one-third of them endowed graphemes with personality or "fondness" (meaning a like or dislike).

But is this really a form of synesthesia? Although the personification of letters and numbers is not encompassed by the narrowest definition of synesthesia—wherein one sense triggers another—we tentatively consider it a form of synesthesia since it represents an automatic, involuntary linking of a learned sequence with another dimension, in this case, the conceptual dimension of personality or gender.[29]

To make the inclusion less tentative, one would have to prove that personification is something more than a cognitive association. Recently, Noam Sagiv and his colleagues in England addressed this.[30] First, they determined that of 248 synesthetes in their database, 32% reported some form of personification (similar to the 35% estimated by Calkins over a century before). When they tested the consistency of the gender and personality descriptions, they found the descriptions were much less consistent than the grapheme → color matches made by the same subjects. This result may weaken the case for grapheme → personality as a genuine synesthetic variant; alternatively, it can mean that consistency testing is not the correct standard for personification, especially since several synesthetes have reported that the assigned personalities can change depending on their mood and stress level.

To examine whether the personifications are automatic, the researchers presented subjects with a new kind of Navon figure: male or female icons built out of letters (see figure 3.12). Over many trials, they flashed such icons on a monitor and asked subjects to decide the gender of the icon as quickly as possible. The experimental idea was to see whether a letter's gender would slow down a subject's response time to an incongruent icon gender. The results indicated a slight slowing in the incongruent case, but the effect was not statistically significant. Sagiv points out that his sample size was small and that different subjects had different confidence levels in their letter-to-gender assignments in the first place. One subject with the strongest associations and the highest consistency showed the Stroop effect the authors were looking for, but the remaining subjects did not.

Thus, whether grapheme → personality should be included as a form of true synesthesia remains debatable, in part because it is difficult to test it as rigorously as grapheme → color synesthesia. Sagiv and colleagues have suggested a few ways in which it could be objectively tested in the future.

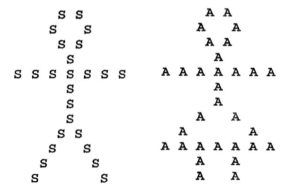

Figure 3.12
When a male or female icon is briefly presented to a synesthetic subject, will the gender associated with the letter slow the subject's response about the gender of the icon?

First, galvanic skin response (a measure used in lie detector tests) could be employed to see whether a greater emotional response results from the presentation of a disliked number or letter. Since the galvanic skin response is rapid and automatic, a positive result would rule out a purely cognitive association. Second, one could use fMRI to see whether a grapheme → personality individual shows increased activation in parts of the brain associated with emotional salience and familiarity with persons (an area called retrosplenial cortex, to which we return in chapter 9). These tests could physiologically verify that some people react differently to letters and numbers than others.

In general, the fact that a third of grapheme → color synesthetes report personalities, genders, or both suggests something nonaccidental about the association. It provides an intriguing neural riddle for future studies to tackle.

Conclusions

Color is the most common expression of synesthesia, with people experiencing hues for digits, letters, and other learned sequences such as weekdays and months. In the next chapter we look at colored hearing, in which sounds trigger color sensations. This implies that in the subconscious landscape, color is a dimension that the brain easily interconnects with compared to other concepts. In chapter 5 we return to another common perception, spatial location, and ask what it is about learned sequences that seems to stick to synesthetic space. For now we are left with the question of why color so easily couples to myriad other concepts, and why those associations—perhaps present in everyone—poke above the surface of awareness in so many people.

4 See with Your Ears

About 40% of synesthetes "see with their ears," according to Sean Day's self-reported survey (see table 2.1). The common name for sound-to-sight synesthesia is colored hearing, meaning the activation of color, shape, and movement by sound. In formal notation, this type of synesthesia is [sound → color, shape, movement, location, . . .]. Triggers include everyday environmental sounds such as dog barks, clattering dishes, voices, and especially music. Dynamically, we liken colored hearing to fireworks because the colored shapes are said to appear, scintillate, and move around, then fade away only to be replaced by a kaleidoscopic montage of colored photisms so long as the varying sound stimulus continues.

In individuals with colored hearing, associations arise between acoustic properties, such as pitch or timbre, and visual qualia (referring to the subjective aspect of perceptual experience such as redness, brightness, or sharpness). Exactly how this happens will become clearer as we explore the similarities among different aspects of sensation as well as anatomical connections between sight and sound areas of the brain.

It is unclear why given individuals respond only to some sounds but not others. Some colored hearing synesthetes are triggered by general sounds, others by sounds that seem musical (e.g., bird chirps), and others only by notes of the musical scale. Even within these categories, not every sound elicits a percept. Some are triggered by the acoustic properties of speech, as in Rebecca Price who described her husband's voice as "a wonderful golden brown, like buttery toast." In contrast, approximately 10% of individuals experience color only in response to the basic linguistic units of speech—that is, to the sound of *phonemes*. Technically speaking, they are categorized as phoneme → color synesthetes rather than individuals with colored hearing.

The most plausible explanation for the broad variety of colored hearing has to do with the different possible time windows during which the genetics unfold, the brain develops, and learning takes place, a topic we return to in chapter 9.

Colored Hearing

Both Rebecca Peacock and Mike Morrow see sound-induced color as a transparent "overlay" on whatever they are looking at. "I'm not sure 'seeing' is the most accurate description," says Rebecca. "I am seeing, but not with my eyes, if that makes sense." Mike Morrow elaborates:

I see shapes and colors in response to sounds. I enjoy electronic music because it evokes such wonderful shapes and colors . . . as if I were looking through a plastic transparency which is in front of my eyes. If I shut my eyes, or if it is at night in the dark, then the shapes are the only thing in the field and are therefore more intense.

However, there is a secondary path. Sometimes when I hear words I will see shapes. This second one is the one that makes me feel silly. You will notice the shape which your last name evokes . . . [see figure 4.1]. I'm not much of an artist. This is the first time I've written something like this down, but it's accurate.[1]

Laurel Smith reports that musical notes appear in front of her visually, in a constant frame (like a picture frame). The music "moves" with her eyes, like the visual aura of a migraine:

I think it is safe to estimate that the entire range of sound I experience occurs within a range of 90 degrees above and below eye-level, with most usual sounds—especially

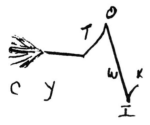

Figure 4.1
Shape of Dr. Cytowic's (misspelled) spoken name, as seen by Mike Morrow.

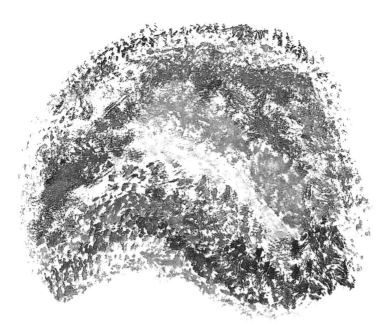

Figure 4.2
For Jane Bowerman, the whoosh of the furnace ignition produces a stack of colored lines.

musical ones—occurring within the range of 45 degrees above and below eye-focus. Human speech normally occurs within about thirty degrees below and above eye-focus.

For Jane Bowerman, environmental sounds set off innumerable synesthesias. The whoosh of a furnace ignition, for example, produces a stack of colored lines like a heap of pancakes viewed from the side (see figure 4.2). Crickets make shimmering maroon circles that ripple out from the center then fade away. The sound of her doorbell looks like gray and brown triangles moving off to the right.

[Sound → color, location] synesthesia has been described in children as young as three[2] but occurs more reliably at ages five to six.[3] There is a delightful case of colored hearing in a three-and-a-half-year-old boy.[4] His color vocabulary is limited to saturated primary colors (e.g., he calls rose and shades of pink "red"). As he was being put to sleep one evening two crickets chirped loudly, one comparatively high and shrill. "What is that little white noise?" he asked. Told it was a cricket, and not satisfied, he said, "No, not the brown one but the little white noise." He then imitated both.

For the boy, an electric fan is orange, a frog's croak bluish, a squeaking door black and white. A small Japanese bell is red when struck loudly, and white when it is faint. In casual conversation he makes remarks such as, "that noise is red, isn't it?" assuming everyone sees what he does. To amuse himself he presses piano keys and tells himself the color of their sounds. Upon seeing a rainbow one day, he exclaimed, "A song, a song!"

A noteworthy characteristic is that the child's palette of colors and light-nesses are arranged in an orderly manner. Middle C is red, with tones below it red and red–purple. Notes further down the bass scale are grey, then black, whereas tones above middle C are progressively blue, green, then white. High tones are lighter than low ones. Although synesthetic sound–sight mappings are idiosyncratic, the mappings are often orderly in this way. This is not surprising given that the brain's primary hearing area is anatomically arranged in an orderly progression. This layout, called tono-topic, means that as you progress along this area of the cortex, you progressively encounter neurons that code from low tones to high tones as illustrated in figure 4.3. We say more about the similarities between synesthesia and normal perception below.

Many synesthetes with colored hearing are said to be *polymodal*, meaning they have multiple forms of synesthesia. For example, MD, the blind synesthete in chapter 2 who sees Braille letters as well as piano and QWERTY keyboards as luminously colored, also sees music as colored shapes:

Violins and similar stringed instruments evoke a nice medium shade of green. Piano music is white, and a piano concerto with a lot of strings in the orchestral accompaniment evokes a green background with white in the foreground. Mozart's clarinet concerto is a wonderfully deep shade of blue, and the music of a flute is red.[5]

Of course the Walt Disney film *Fantasia* is built on the idea of sound-to-sight correspondences, and many synesthetes say it is a reasonable illustration of what it is like to see sound.

Another synesthete, Peter T, perceives purple triangles when he hears certain chord progressions, and he sees stars—very yellow ones—whenever cymbals crash. Suzanne Y experiences a purple herringbone pattern when listening to Bach. Not knowing there was a name for synesthesia, she used to call it "spatialization of sound." Because associations are idiosyncratic,

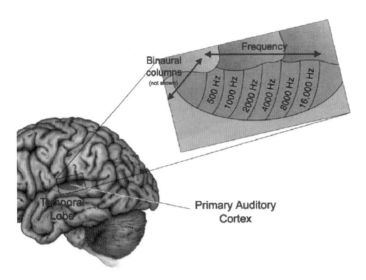

Figure 4.3
The auditory cortex, located in the temporal lobe, has an organized layout of tones
from low to high.

synesthetes often regard each other's experiences as absurd, insisting, for
example that, "Only music in the key of F is yellow, and it's harvest gold
at that."

The previous examples illustrate some of the acoustic and musical properties that synesthetes typically respond to: sounding pitch, pitch class,
musical key, timbre, chords, melody, and volume. No wonder colored
hearing is such a rich experience. To clarify some of these terms, "pitch"
refers both to whether a note is high or low (pitch height) and what it
specifically is (e.g., C, B♭, F♯, etc., this latter distinction being called "pitch
class"). Music in a given "key" means it adheres to one of the major or
minor scales, whereas "timbre" is the distinct quality of tone that lets you
distinguish, say, a violin from a flute playing the same note at equal loudness. (The German word for timbre is "*Klangfarbe*," literally "tone color.")
"Chords" are two or more notes played simultaneously, and here the intervals between notes determine the synesthetic response. "Melody" refers to
notes that are arranged sequentially. Any or all of these factors can determine the exact nature of colored hearing.

The early literature reports two interesting subjects who sense both color
and three-dimensional shapes in response to tones.[6] In these subjects

Table 4.1
Shapes experienced in response to musical tones

Instrument	Subject A	Subject B
Flute	Thimble or acorn cup	Hollow tube
Saxophone	Cup with solid inner core	Bursting of a mass into rough, jagged, splintery particles
Harmonica	Spatially distributed discs	Flat rectangle
Jazz whistle	Thick waving streamer	Lumpy dough–like elongated mass
Musical saw	Elongated globule, jagged surface	Yards of round ribbon–like material
Cello	Flat horizontal base with spring-like vertical protrusions	Thick ribbon
Violin	Tubes with enlarged nodules	Ribbon much thinner and smaller than cello
Piano	Quadrangular blocks	Spheres

(From Ziegler [1930].)

synesthesia occurs chiefly when hearing solo renditions of various instruments, and emotionally it possesses "high aesthetic appeal." In both subjects, pitch determines hue and lightness whereas timbre affects the shape (see table 4.1). "Every instrument excites a specific form, which maintains roughly the same features at all pitches, intensities, and durations." Given the varieties of synesthesia, however, it should not be surprising that different cases exist wherein both pitch and timbre affect color.

Laurel Smith is a deeply synesthetic musician who experiences colors in response to music, speech, and other sounds. Musically, her colors are determined by pitch, timbre, sound structure, key, counterpoint, dissonance, and musical style. At the simplest level, her pitches are experienced as colored lights (see table 4.2). Laurel Smith's colors for pitches are based on the diatonic scale. Sharps and flats are seen as variants of their natural counterparts.

She comments that if a note is out of tune, it has a white halo if it is sharp and a dark halo if it is flat. Flats and sharps based on a key appear without halos. Accidentals foreign to a key are seen as "sharp" or "flat." A silverish halo indicates their foreignness to the key.[7]

For Laurel, pitch height affects the shade of a given sounding note, with higher sounds being lighter and lower ones darker. For example, the goldish

Table 4.2
Colors in response to pitches in the diatonic scale

A = Pink
B = Blue
C = Goldish–white, like sunlight
D = Silvery white, like moonlight
E = Fiery orange
F = Tree bark brown on written page; purple when sounding aurally or in head
G = Green, like leaves

tint of C darkens as it descends the keyboard. D takes on more silver–gray overtones as it descends until it becomes almost black. She also experiences pitch as sized: lower notes are bigger than higher ones, especially bass notes, which are often two to ten times as big as treble notes. Bass notes are also more shapeless.

Laurel is quite sensitive to an instrument's timbre. Often, several sounds fall under a single color. For example, the following are all purple: the viola's D and A strings, the cello's C and D strings, a piano's third octave from the bottom, and the gadulka (a Bulgarian string instrument). She warns that such color associations are grossly oversimplified because timbre is even further differentiated by the "sound structure" of a particular instrument. "It is as inconceivable to hear a flute without seeing an intensely bright pitch color as it is to look into a well-polished mirror and see nothing."

Franz Liszt and Nikolai Rimsky-Korsakov were said to have disagreed over the colors of musical keys. In 1842 when Liszt took over the post of Kapellmeister in Weimar, he astonished the orchestra by saying, "Oh please, gentlemen, a little bluer, if you please! This tone type requires it!" Or, "That is a deep violet, please, depend on it! Not so rose!" The orchestra eventually got used to the maestro seeing colors where they saw only notes.[8]

It is perhaps not surprising that several famous musicians have colored hearing. Synesthetic composers besides Liszt and Rimsky-Korsakov include György Ligeti (whose music was used for the soundtrack of *2001: A Space Odyssey*), Amy Beach, Jean Sibelius, and Olivier Messiaen. Modern musicians include violinist Itzhak Perlman, oboist Jennifer Paull, jazzmen Michael Torke, Thomas Wood, and Tony De Caprio, and pop artists Eddie Van Halen and Stevie Wonder.

The French composer Olivier Messiaen[9] is particularly interesting not only because his synesthesia is bidirectional—music is color and color is musical—but also because he invented a method of composition specifically to convey music in color terms. He called it the modes of limited transposition and it is so stylistically unique as to make Messiaen's music instantly recognizable. Mode 2, for example, is a certain shade of violet, blue, and violet–purple, whereas mode 3 is orange with red and green pigments, spots of gold, and a milky white with iridescent reflections like opal. Consequently Messiaen speaks of "color chords." The modes are not harmonies in the usual sense, nor are they even recognizable chords. "They sound like colors," he says flatly. To speak of an exact correspondence between a key or a conventional chord and a color is not possible because his colors are complex—he often speaks in terms of a stained-glass-window effect—and they are linked to equally complex sounds.

Whenever Messiaen *hears* or *reads* music he sees colors; conversely, he frequently spoke of translating colored landscapes into music.[10] For example, his 1977 symphony *Aux Canyons des Etoiles* ("From the Canyons to the Stars") was famously inspired by Bryce Canyon, "the most beautiful thing in the United States. The piece I composed about Bryce Canyon is red and orange, the color of the cliffs." Take the following attempt at color-to-sound translation. As the intensely blue Steller's jay flies over the canyon,

his belly, wings and long tail are blue; the blue of his flight and the red of the rocks takes on the splendor of Gothic stained-glass windows. The music of this composition attempts to reproduce all these colors.

For the Steller's jay, chords with "contracted resonance" (red and orange) . . . Chords with "transposed inversions" (yellow, mauve, red, white, and black) render the colors of the rocks. . . . Next, polymodality superimposing the three 4-mode (orange-colored with red strips) to the six 2-mode (brown, reddish, orange-colored, purple) brings to a fortissimo conclusion the sapphire blue and orange red rocks.[11]

Scientific accounts of phenomena should be predictive. Accordingly, Princeton musicologist Jonathan Bernard has demonstrated the correspondence between color and sound structure in Messiaen's work through conventional musicological analysis.[12] He finds that color is predicted by the vertical spacing of notes. That is, chords formed by modal transpositions have characteristic spacing, and two different spacings of the same modal set correspond to two different colors.

To Messiaen, there is no such thing as a single note. He is always aware of overtones and harmonics, especially in natural sounds such as wind, waterfalls, and birdsong that so often populate his compositions. Where we might hear one sound, Messiaen hears many sounds within it. What we hear as a single chord of pitch-class notation 2, 2, 2, 7, 8, 6, 4, Messiaen hears as multiple sounds. Conversely, inasmuch as his synesthesia is bidirectional, we might perceive one color whereas Messiaen sees all the nuances so typical of synesthetes describing colors "just so."

Even though he had been composing with the modes before age 20, Messiaen's first public mention of synesthesia was a passing reference in *The Technique of My Musical Language* to "the gentle cascade of blue–orange chords."[13] His associations are involuntary and consistent, suggesting that the relevant acoustic properties are not dependent upon the sonic attributes of a given performance.

What separates Messiaen from most persons with colored hearing is that he discovered particular sound combinations that evoke a variegated range of colors, allowing him to "paint the visible world in sound." Messiaen sees three types of colored sounds. The first is monochromatic, labeled simply "green" or "red," for example. The second sound type is a two-colored mixture described in hyphenated names such as blue–orange. The third is a more complex mixture of pairs ("gray and gold"), triplets ("orange, gold, and milky white"), or a dominant color that is "flecked, striped, studded, or hemmed" with one or more other colors. The primary evidence for Messiaen's sound–color combinations comes from biographers,[14] from copious notes that Messiaen wrote about his compositions, and from color notations printed on the published scores themselves.

Perfect Pitch

Synesthetes often ask if there is a relationship between perfect pitch and synesthesia. Perhaps it makes intuitive sense that assigning specific colors to given musical pitches would make it easier to remember absolute pitches later on. So far as we know, however, there is no increased likelihood that having one trait means having the other. The two traits are separate, although what is most analogous between them is that each has a strong genetic component.

While most people understand that physical traits such as blue eyes or red hair are inherited, they do not consider that psychological or perceptual traits may be as well. For example, we know that single genes are important in varieties of both hereditary deafness and blindness. Little work has been done on the sense of smell, although scattered reports indicate hereditary factors at work in selective anosmia—the inability to detect specific odors.[15] Conversely, the skill of talented "noses" in families of parfumeurs and cognac industrialists is legendary. Of course, one talent that shows a strong tendency to run in families is musical ability. The pedigree of Johann Sebastian Bach is a prime example, though hardly unique.

Like synesthesia, perfect pitch runs in families and manifests at a very young age. [16] Women with this trait outnumber men.[17] Perfect pitch is similar to synesthesia in four aspects: (1) It is all-or-nothing; the talent is either present or it is not. (2) The skill appears naturally without the necessity to develop it through the practice that characterizes the acquisition of other musical skills. (3) Most individuals with perfect pitch are astonished to learn that not everyone has the capacity. (4) It manifests at an early age: 26% by 5 years old and 89% by 10 years old. Pedigrees are compatible with autosomal recessive inheritance, and the fact that 80% of affected individuals are female suggests that sex-related factors influence its expression.

Joseph Long is a concert pianist who has both synesthesia and perfect pitch. He demonstrates how both operate independently and regards the latter skill as one of "learning to label" at a critically early age:

I started playing the piano at age four on my grandparents' battered old instrument, which (I later found out) was tuned a minor third or so below A-440 [the concert-pitch standard]. I had synesthetic colour associations from the very first session onwards—C was blue, D was green, etc. I have clear memories of childhood, and am confident that this account is accurate.

At the age of five, my parents bought me my own piano, which was tuned a little higher. I didn't know then it was a tone below A-440—it just sounded higher to me. I then enrolled for piano lessons with a local teacher, and her piano was—guess what—bang on A-440. I was now playing three pianos, all at completely different pitches.

The important thing at this stage was that, on all three pianos, C was still blue, D was green, and so on, regardless of the sounding pitch. I just didn't figure at the time that the sounding pitch, which was radically different on all three pianos, was all that important. It could be discarded as irrelevant. Middle C was the physical key I pressed, not the sound—that was what everyone seemed to be teaching me.

And then a turning point happened that led to absolute pitch becoming one of the most important parts of my musical life. It came in the form of a visit from my friendly piano technician after my fifth birthday when I had been practicing a good many months. He immediately said it needed a pitch raise up to A-440. Once he carried out the work it sounded like my teacher's piano. So it was then that I grasped that pitch was "absolute," or at least that there was such a thing as standard concert pitch at which my teacher's piano had been set and at which the others had not been. Evidently something in me before then was able to hear and remember distinct pitches even though I had associated colour with the physical piano key rather than the sound it made when pressed.

That, at any rate, was the situation for Joseph when he was young. Over time it has changed as learning continues to influence his perception such that as an adult the two phenomena—synesthesia and perfect pitch—are no longer distinctly separable. In fact, Joseph says, "it is as though my absolute pitch has hijacked my synesthesia." What he means by this is that he must make a mental adjustment if the sounding pitch of, say, a recording does not match his synesthetic color for a given key signature. For example,

my colours will act as labels for the closest I can find to the absolute pitches I'm hearing. If a recording is at Baroque pitch (A–415, i.e. a semitone lower) then the color of a piece in B flat will be orange (i.e. the colour I would normally associate with A). It will definitely not be the sort of dark pink and black-and-white flecked with various other and interesting strange shades that is my color for B flat.

What this shows in Joseph's case is that ever since his youthful "turning point," the acoustic properties of a physical sound have gradually outweighed his conceptual categorization of what, for example, it means to be in the key of B flat. Most often in synesthesia, as we've seen in the case of colored graphemes, it is the concept rather than the lower perceptual parameters that drives color (as we showed by the fact that capitals, lowercase, and different fonts usually all result in the same synesthesia). Joseph

Long's experience further illustrates that the synesthete's brain can be dynamically plastic even in adulthood.

Laurel Smith, who also has perfect pitch and synesthesia, likewise speaks of "transposing" her colors depending on the sounding pitch. For example, "when I was a child I found I could play on a piano tuned half a step down by changing the colors I saw on the keys (perhaps this is how I could later play a Baroque instrument)." Similarly, having first learned her colors for the piano's treble and bass clefs, she later took up the viola with its alto clef by "practicing seeing the correct colors on the clef." As we noted, Laurel has colors for musical timbre that differ from her pitch colors. With orchestral music, and even non-Western music such as the Javanese gamelan or the Japanese shakuhachi that are tuned to unique scales, she automatically sees pitch colors unless she deliberately shifts her attention to timbre color.

There is thus in the sense of hearing a greater interplay between perception and conception than there is in vision.

Beyond Color

The twentieth-century modern artist Wasilly Kandinsky left behind sufficient writings to indicate he was synesthetic. Listening to Wagner's *Lohengrin*, for example, he said: "I saw all my colors in my mind; they stood before my eyes. Wild, almost crazy *lines* were *sketched* in front of me." "Wild, almost crazy lines" might easily describe Kandinsky's own abstract canvases. He combined four senses synesthetically: color, hearing, touch, and smell. He said color had touch and texture: orange, for example, was prickly, whereas dark ultramarine was smooth and felt like velvet. Colors also had distinct smells. He said, "You know there is a common expression, the 'scent of color,'" which probably no one thinks is common, but it says something about what went on in Kandinsky's mind.

Beyond having color and texture, sounds can also be synesthetically linked to gender and personality (just as graphemes are). For Laurel Smith, the pitch of G-sharp is

Feminine. Energetic, emotional, imaginative, and outgoing. Somewhat ethereal and detached, even while interacting with others. "Different." Curiously cool, calm, lively, and hyperactive all at the same time. Seems to float above everyone else, her different way of perceiving the world around her enticing her higher and higher with secret delights only she can see.

She also perceives gender and personality with instrumental timbre (see table 4.3). Like Kandinsky, Laurel also experiences shapes for given sound structures:

Trumpets have more vertical lines than French horns. Electronic sine waves are entirely flat, smooth, and undifferentiated . . . so bright and shiny as to be hard to look at for long. It is in aperiodic noises such as percussion that sound structure is most obvious. Many have vertical attacks and extended, broadly sloping decays.

Finally, Laurel experiences a rare audiomotor synesthesia in which sound triggers a sensation of movement, or of striking a particular posture. For example, hearing a particular harmony in the bass line makes her body feel as though it is moving in ways described in table 4.4. We thus see that while sounds most often induce color, the variety of synesthetic experiences in response to it is as wide as those observed in other forms of synesthesia.

Phonemes as Colored

In the previous chapter we explored colored graphemes, emphasizing that for most synesthetes the shape or concept of the letter determines its color. But about 10% of synesthetes are sensitive to the auditory component, or phoneme. For synesthetes such as Laurel, the color of e differs depending on how it is pronounced in a word. For her, long vowels are always lighter and brighter and are luminescent shades that move and seem transparent. She experiences colors not only for all phonemes found in the English language but also for phonemes found in other languages (as in table 4.5). Knowing several languages, Laurel recognizes phoneme–color similarities: "Vowels that induce palatalization, such as English long u, and most Slavic e's and i's or Brazilian final e's have yellow borders on the leftmost sides, the same yellow as y, the sound they are most like."

The coupling of colors with phonemes can sometimes lead to confusion when subjects are asked whether their letters are colored: after all, a synesthete's colors may be entirely driven by the sounds of phonemes but can

Table 4.3

Genders and personalities of instrument sounds for Laurel Smith

Viola	*Feminine* Warm and deeply caring. A deep thinker and feeler, philosophical and thoughtful. Wise and quite experienced. Gentle and loving, affectionate. Usually hugs the listener. Can become quite melancholy. On C-string, rather nervous and high strung.
Cello	*Feminine* The ultimate mother. Deeply loving, caring, generous, and kind. Extremely protective of those she loves. Always hugs the listener, especially on the D- and A-strings. Deeply emotional, compassionate, and empathetic.
Double bass	*Masculine* Stable as a rock and strong, yet deeply empathetic, warm, and caring. Can be quite tender-hearted. The ideal father and strongly hinting of God, the Savior, and other religious Greats such as the Dalai Lama. Often hugs or otherwise physically demonstrative towards listener.
Flute	*Feminine* Sweet, innocent, and naive with an artless, childlike simplicity. Prone to great frolicking energy but also often profoundly pensive and surprisingly insightful. The ultimate in bird-likeness. The A-natural of the instruments.
Oboe	*Masculine* Profoundly emotional and thoughtful. Withdrawn, introspective, and prone to melancholy.
Clarinet	*Masculine* The clarinet is the nerve barometer of the orchestra. It is the most profoundly calm and serene in certain circumstances and the most nervous in others. In the clarino register, he is extremely easygoing. Capable of profound calm and sedateness, though rarely in a truly mystical fashion. The clarino register is quite down-to-earth, realistic, and happy-go-lucky. In the chalumeau register, can still be quite calm at slow tempos, but can often be quite nervous, especially in faster tempos and runs or in minor keys.
Bassoon	*Masculine* Highly intelligent but profoundly awkward. The proverbial academic nerd. Thoroughly sentimental and romantic. Always yearning for his beloved, the bassoon is driven to distraction. This leads to many amusing moments of awkwardness and humor. A good-natured fellow with a quirky but enjoyable sense of humor. Not particularly sociable but not particularly withdrawn. A bit self-conscious as well as sincerely humble in a touching way.

Table 4.4
Laurel Smith's kinetic positions and motion sense caused by functional harmony (bass line)

I—Tonic—Standing upright with feet on ground or floor.
II—Supertonic—Flying low above the ground.
III—Mediant—Stepping on a stair.
IV—Subdominant—Soaring high in the sky.
V—Dominant—Poised on a landing with one's knees bent ready to jump.
VI—Submediant—Floating around in stratosphere (very little gravity).
VII—Leading Tone—Tripping onto the top step of a flight of stairs.

Table 4.5
Colored speech for "irregular" sounds and sounds in other languages

Affricativized C [ts] = Sunlightish white with the yellow of t (German z, West Slavic and Hungarian c; Japanese tsu)
Bilabial F = Green as light as j but not so brilliant (Japanese f)
Centralized vowels = Candle-flame yellow; generally includes reduced vowels in time-stressed languages as well as sounds such as English oo in "book" or "look" and German or Hungarian ö
Palatalized S = Deeper, solid sky blue, also for "S" in English "Sh"
Palatalized Z = Yellowish tint (z as in "azure")
Gutturals, [x] = Same as k except for sound structure, the voiced version being the same color as g but darker and with more complex sound structure
Ch/palatalized t's = Sunlightish, between t and c, but with the sound structure of its constituents

be mistaken for grapheme → color merely because subjects are reading aloud in their head. For that reason, it is important to clarify this distinction when interviewing synesthetes. Typically, a synesthesia dependent on sound is sensitive to the difference between homographs such as "wind" ("How did I *wind* up here" vs. "The *wind* blew me here").

Going in Both Directions

For some synesthetes, colors induce sound just as sound induces color. Wassily Kandinsky claimed that each color had an intrinsic sound, a relationship he elaborated in his 1912 book, *On the Spiritual in Art*.[18] He later attempted to equate color with Schoenberg's twelve-tone music. In other words, Kandinsky increasingly intellectualized his synesthesia, deliberately turning it into an abstraction. He sought a universal translation among the senses that he hoped would apply to everybody (even though we now know that cannot work because synesthetic associations are idiosyncratic).

The rare times that synesthesia goes in both directions usually involve sight and sound. Although some people like Marcia Smilack have no problem with it, others find bidirectional synesthesia overwhelming at times.

Julie Roxburgh is a British music teacher with colored hearing who has been studied extensively by Simon Baron-Cohen and colleagues to verify that her synesthesia is highly consistent in both directions.[19] Julie sees colors when she hears sounds, and hears sounds when she sees colors. As she takes in a visual scene, each color produces a different musical note, while any speech or environmental sounds she hears also trigger their own colored photisms. The onslaught of cacophony results in considerable perceptual interference and causes her distress. She copes by leading a relatively restricted life in the country and avoiding both loud colors and noisy environments (both visually and acoustically). For the documentary, "Orange Sherbet Kisses," the BBC filmed her walking in Piccadilly Circus at night while navigating its traffic and neon signs:

This is an area I avoid if I possibly can. Every one of my senses is being battered. I find it very difficult to keep control because I'm not quite sure whether what I'm seeing is what I'm hearing or what I'm hearing is what I'm seeing. I find it difficult to avoid the traffic, avoid the people. The lights themselves are creating sounds. There is a flashing light that also gives me a tactile sensation in my fingers. The color green of the little man on the cross sign is screaming a horrendous yellow at me. Behind that are neon lights which are shouting . . . it's like having nails in the back of my throat. . . . It makes me feel frightened, tired, exhausted. I'm losing control. I don't think I can stay here very long at all.

While synesthetes often speak of having a "gift," Julie Roxburgh's distressing sensory confusion indicates that synesthesia can be maladaptive in a small subset of individuals. It raises the question of whether there are competing forces at work: a possible creative advantage in maintaining synesthesia past childhood versus the disadvantage of perceptual confusion when retained synesthesia is too intense.

The problem for synesthetes who are overwhelmed seems not to be bidirectionality *per se* but instead the intensity of the unusual channel of vision to sound. For example, Lidell Simpson's synesthesia goes overwhelmingly in that direction. That is, he hears whatever he looks at, especially if it moves or flashes. Lidell was born profoundly hard of hearing. "I can turn my hearing aids off, but true silence I have never known," he affirms, because his vision continues to provide sound. Thanks to bilateral hearing aids, he hears sufficiently well to have learned several foreign languages. He says,

I also hear with equal clarity other things that are not "sonic." . . . Photonic hearing to me is the result of light. My eyes are another pair of "eardrums" to me. Every color "emits" a tone. Intensity, brightness, position—all influence the "tonal" quality of these emissions.

For example, there is a radio tower miles in the distance. On the towers are a series of lights, red or white (each color has its own "note," "tone," or "key" if you will). I hear the blinking of the lights and its intensity increases as I approach. Now add the reflectors along the side of the road. Every one of them I see emits its "ping," and the center striping of the road emits its own sound. Every car headlight has its tune. The tonal quality changes with respect to relative position, like the Doppler effect.

Even in the daytime, same thing. I hear the sky, the trees, anything my eyes perceive emits sound.

Lidell is also polymodal, touch evoking sound as well. Alcohol amplifies his synesthesia. He says a few beers in a quiet setting are manageable, "but at a very noisy bar I am courting disaster."

How Much Synesthesia Is Normal?

If we can show that synesthetes and nonsynesthetes both map sensory dimensions similarly, we can infer they share some perceptual processes, and so perhaps synesthetes use some of the same neural pathways employed in normal cross-sensory perception rather than having a special "short circuit" between brain areas serving sight and sound. Three decades ago Lawrence Marks[20] did just that, affirming that there are straightforward relationships between pitch, lightness, size, and loudness as well as visual position and shape in both nonsynesthetic and synesthetic matches (see figure 4.4).

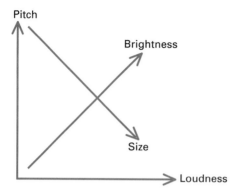

Figure 4.4
Both synesthetes and nonsynesthetes map pitch, size, brightness, and loudness in an orderly and lawful way.

Dr. Marks reminded us that the relationships between sensory dimensions belonging to different senses are orderly and lawful. For example, both synesthetes and nonsynesthetes say that loud tones are brighter than soft tones, that high tones are louder, brighter, and smaller than low tones, and that low tones are both larger and darker than high ones. Melodic intervals also map to bright–dark values: lighter stimuli are said to go with ascending melodic intervals and darker stimuli with descending ones, whereas larger melodic intervals produce more extreme (lighter or darker) choices than closer intervals.[21]

Recently, Jamie Ward and colleagues compared synesthetes and nonsynesthetes directly for the first time[22] by playing a series of musical notes and asking both groups to pick their color choices from a computerized palette that offered both preselected and customizable colors. Synesthetes chose significantly more customized colors than controls, and both groups showed an identical pattern of systematically mapping low pitches with dark colors and high pitches with light colors. Dr. Ward further showed that synesthetes were far more consistent than controls in their choices over test–retest sessions (figure 4.5) and that they generated their colors automatically whereas controls had to deliberately decide what color best "went with" a given tone. Automaticity was demonstrated by cleverly inducing cross-sensory Stroop interference while naming a color patch that is incongruent with the color of a simultaneously presented tone (e.g., a "red" tone interferes with naming a blue color square).

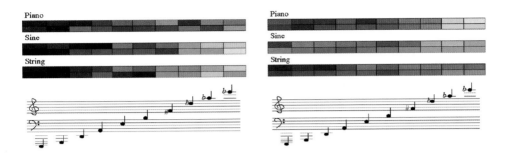

Figure 4.5

Colors chosen on two occasions (stacked bars) for ten single piano, sine wave (pure tone), and string timbres for a control subject (*left*) and a synesthete (*right*). By comparing the pairs of bars, note the far greater consistency in the synesthete and the matching of higher notes to lighter hues. For this synesthete, piano and string timbres are literally more colorful than pure tones generated by sine wave. (From Ward et al. [2005], with permission.)

Experimentally, Dr. Ward showed that two notes played simultaneously (dyads) elicited two or more colors that differ from those of their component notes. This is consistent with what Messiaen tells us and with the experience of a teenage boy born with impaired vision (20/100) whose chord colors are not the same as the individual notes that make up the chord.[23] This indicates that interval rather than pitch determines colors in these particular synesthetes. The reason pitch can produce multiple forms of synesthesia is that it is a multidimensional attribute.

Are We All Silently Synesthetic?

Sound-to-sight synesthesia is particularly interesting because it may be present in all human infants. Multiple lines of evidence suggest so.[24] If true, then adult synesthetes possibly retain juvenile circuitry or physiology that most individuals lose as they mature.

Tradition has viewed perception as modular, meaning that sensory channels are separate, with little, if any, interaction among them. The modular concept unfortunately masks the existence of rich multisensory interactions in the normal brain.[25] In everyday encounters, we do not experience sensory events in isolation because each sense receives correlated input from other senses about the same object or event. Unnoticed by the perceiver, each sense modality is highly influenced by other senses.[26] In adults, for example, sight, sound, and movement influence one another so closely that even bad ventriloquists convince us that the moving dummy is doing the talking. The ventriloquist effect also occurs while watching movies when voices are perceived to originate from the actors on screen despite the sound coming from the surrounding speakers. The illusion is compelling and takes place without the need for attention. Even the very young perceive it.

Another example of the influence vision has on hearing is the McGurk effect,[27] in which the sound of /ba/ is perceived as /da/ when coupled with visual lip movement associated with /ga/. The McGurk effect demonstrates that voice and lip-movement cues combine early on, before the audio and visual signals are assigned to a phoneme or word category.[28]

Although vision typically dominates hearing when the two compete, sometimes the relationship goes the other way. Take the illusory flash effect: when a flashing light is accompanied by two beeps, it appears to flash twice.[29] A related illusion is called auditory driving, in which the apparent speed of a flickering light seems faster or slower depending on the rate of an accompanying sound.[30] Simple illusions like these powerfully reveal that vision and hearing are tightly coupled at very early levels neurologically.[31]

Recently, new techniques have been applied to cross-sensory perception. Electrode recordings from single cells in the brain show that when a bang and a flash occur at the same time and location, the activity level of the cells can increase to a level exceeding that predicted by adding up the responses to the single-sense inputs.[32] And neuroimaging studies verify that seen speech influences heard speech at a very early stage in the brain (as in the above McGurk effect).[33] In another example, touch influences what we see by way of very low-level connections in the brain.[34] Finally, even if we do not consciously perceive a facial expression, it colors our perception of emotion in the speaker's voice, and electroencephalography (EEG) indicates that this integration happens at very early stages.[35]

Remarkably, four-year-olds agree with adults as to the correct cross-modal matches among loudness, brightness, and pitch as illustrated in figure 4.4. Even one-month-old infants equate certain levels of brightness to loudness. Moreover, the congruencies are demonstrably more hardwired than dependent on context. The fact that such links are established early[36] suggests that perceived similarities are not related to language. Because adult perception does utilize context, we can therefore infer that invariant cross-sensory correspondences exist in infants but change during development to become primarily contextual.[37]

An obvious conclusion from the foregoing is that sensory interaction occurs routinely. The two prevailing theoretical views of how multisensory influence develops are known as "integration" and "differentiation."[38] The integration view proposes that distinct sensory channels are initially separate after birth, gradually becoming integrated during the course of development and repeated experience with the world. The differentiation view holds the opposite position, that the various senses form a primitive unity early in development and differentiate from one another as the infant matures.[39]

Evidence shows that a variety of mammalian newborns, including humans, have working anatomical connections between sensory areas. The ones between sound and vision are especially robust. For example, sound evokes recordable visual responses in human newborns. These decrease after six months but are still detectable between twenty and thirty months.[40] The macaque monkey, our close relative, appears to have lifelong connections to visual area V1 from both primary auditory cortex and multisensory areas of the temporal lobe.[41] Because forming such connections does not depend on visual input,[42] their existence may be genetically programmed.

Human infants are skilled perceivers of nonmodal properties such as intensity, rate, duration, rhythm, time synchrony, and location in space. For example, the sight and sound of clapping hands share time synchrony, action tempo, and a common rhythm. The infant's adept detection of nonmodal relations is a fundamental component of selective attention. The detection of nonmodal information at a young age supports the view that development *proceeds from a global unity of the senses to increasing differentiation*.[43] Mounting evidence that the separate senses are not so separate therefore highlights the importance of differentiation in early development.

Rather than a competing dichotomy, however, integration and differentiation are best considered complementary. Differentiating nonmodal relations *and* integrating modality-specific inputs across the senses interact with each other during the early development of perceptual and cognitive skills.

There is much evidence, therefore, that perception is fundamentally multisensory, yet we are rarely aware of the full extent to which this is true. If it is overtly so in infancy while differentiating away as the brain matures, we can then propose two possibilities for why certain individuals remain synesthetic. Either they retain more of the juvenile interactions that most individuals lose, or else they explicitly draw on normal multisensory processes that have grown implicit in the majority. We pursue the view that synesthetes use network pathways that ordinarily underlie the normal integration of multisensory input. An extension or exaggeration of existing pathways eliminates the need for synesthetes to have wholly novel neural structures compared to nonsynesthetic individuals.

Conclusions

Hearing and vision are tightly coupled in the brain as demonstrated by a variety of illusions such as ventriloquism and the McGurk effect. These illustrate that hearing and vision influence each other at extremely early stages. In most people this communication is below the level of conscious access. But for some fraction of the population, the coupling of hearing and vision is explicit. For these synesthetes, music, speech, noise, or phonemes can trigger extraordinary displays of color and light. In chapter 9 we delve more deeply into the neural basis of this form of synesthesia.

5 November Hangs above Me to the Left

For most of us, February and Wednesday do not occupy any particular place in space. However, some synesthetes experience precise locations in relation to their bodies for numbers, time units, and other concepts involving sequence or ordinality.[1] They can point to the spot where the number 32 is, where December floats, or where the year 1966 lies. Like all synesthetes, they express amazement that not everyone visualizes sequences the way they do. These objectified three-dimensional sequences are commonly called "number forms," although more precisely the phenomenon is called "spatial sequence synesthesia." Formally, we note it as sequence → location.

In reality, many of us possess implicit number lines for sequences. When asked, we might agree that the number line for integers increases as one goes from left to right. Spatial sequence synesthetes differ in that they experience sequences explicitly in three dimensions as automatic, consistent, and concrete configurations. The forms have a location in mental space that can be precisely pointed to. If you are not a spatial sequence synesthete, imagine your car parked in the space in front of you. Although you do not physically see it there, you will have no trouble pointing to the front wheel, the driver's side window, the rear bumper, and so on. The car has three-dimensional coordinates in your mental space. So it is with automatically triggered number forms. In fact, even blind subjects can experience number form synesthesia.[2]

Number forms are often panoramic and dynamic, meaning that the viewer can zoom in and out and "move around" within the configuration as they change viewing perspective. Even as viewers look right or left, "up to" or "down on" the configuration, relationships between elements within the form remain constant. Quite often, changing the viewing perspective

Figure 5.1
Marti Pike's days of the week form a simple left-to-right string. Holidays like Thanksgiving and appointment days become "marked" with a hump so that the day is "larger and spot lighted . . . and there is no need for my writing it down." (From Cytowic [2002], with permission.)

alters the "illumination" as if a spotlight somehow highlighted the immediate viewing area. Some visualizers experience their forms as colorless, others perceive them in color.

The most common type of spatial sequence synesthesia is a form for days of the week (see figure 5.1) followed by months of the year (see figure 5.2), the counting integers, and years grouped by decade. In addition to these common types, we have encountered spatial configurations for shoe and clothing sizes, baseball statistics, historical eras, salaries, TV channels, temperature, and more. Some individuals possess a form for only one sequence; others have forms for more than a dozen (see table 5.1).

Do number forms really count as a form of synesthesia? Although at first glance they may not seem to fit the definition of synesthesia as a *sensory coupling*, number forms are considered synesthetic because they unite different conceptual properties. In this case, the concept of a sequence triggers an experience of location in three-dimensional space. Shape, texture, color, and illumination can also be bound to the form. Listen to Marcia Smilack describe her days of the week as

rectangular blocks, upright in a row like dominoes angled about 25 degrees. I see them to my right in peripheral inner vision. Each month has a shape and a color, and these lie on an oval. . . . Starting with January at the bottom, they are arranged counterclockwise so that February (light green) is to the immediate right of January while December (royal blue) is to its left. At the top of the oval are August, July, and June. Each month has a unique shape and size [see figure 5.3].

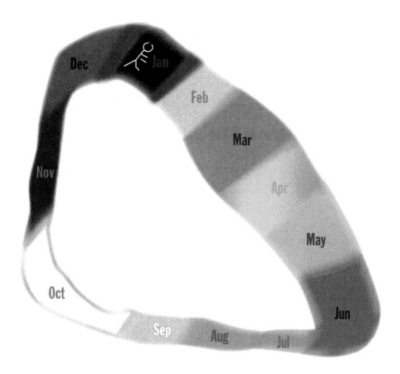

Figure 5.2
Pat Duffy's calendar form. It is common for the months to be unequally spaced. Most calendar forms are circular, elliptical, or horseshoe shaped. (From Duffy [2001], with permission.)

Table 5.1
Sequences seen by Colleen Silva

Numbers	Days of the week
Months of the year	Body measurements
Shoe sizes	Weight
Height	Temperature
TV stations	Geographical maps
Body temperature	Multiplication tables
Time	Alphabet
Grades—a grading scale and pattern for grade-point averages	
My life ⎫	
My ages ⎬ all different patterns	
My school ⎭	

Figure 5.3
Marcia Smilack's calendar form for the months.

In truth, the simple definition of synesthesia as cross-*sensory* pairings has always been inadequate. The case of colored graphemes, for example, does not even involve two different senses given that written letters and colors are both visual. A more comprehensive definition of synesthesia would go something like this: Synesthesia is a hereditary condition in which a triggering stimulus evokes the automatic, involuntary, affect-laden, and conscious perception of a physical or conceptual property that differs from that of the trigger. Under this definition, the conceptual coupling of weekdays and space (and often color as well) qualifies as a form of synesthesia.

Varieties of Timelines and Sequences

Marti Pike's calendar, alphabet, and integer forms were illustrated in figures 2.4 to 2.6. When she thinks of a particular month, the view "fades to the numbers 1–30 (31)," meaning the calendar days. If you inspect figure 2.4, you can see that the serpentine-shaped form for the days of November inherits the overall brown color of November when it is conceived of as a month. Nesting one form within another like this is a typical feature for her. When Marti thinks of or plans out an individual day (see figure 5.4), her spiral-shaped twenty-four-hour clock comes into view. She says,

I see the numbers as though I were below and to the left of 6. Starting with 1, the day *spirals upward* to 12, that is from 12 midnight to 12 noon *above it* and upward

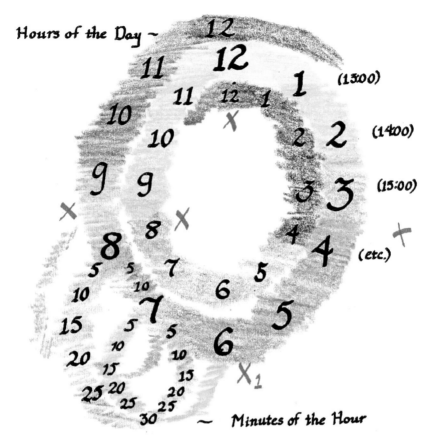

Figure 5.4
Marti Pike's overlapping three-dimensional spirals for hours and minutes of a given day.

to 12 midnight again. For 12 to 6 or 7 a.m., it is gray–black, from 6 on it gets lighter until at noon it is blindingly bright (yellow–white), then gets darker between 5 and 7 p.m. Eight o'clock to ten is a soft blue and from ten on is dark blue.

Her form for the minutes of the hour nests within her clock spirals (see figure 5.4):

For a single hour I see a close up of that hour like a regular clock, only with a "2" in place of the "12," etc., or for the minutes, "20" in place of the 4 . . . Again, I see time as space. Digital clocks drive me crazy!

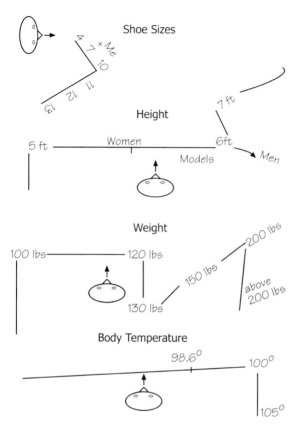

Figure 5.5
Colleen Silva's shoe sizes, height, weight, and body temperature patterns.

Like Marti Pike, Colleen Silva experiences forms for many concepts. Table 5.1 listed eighteen of them, whereas figure 5.5 illustrates a few of her configurations. "Whenever I think of a shoe size larger than 10, I imagine the tip bent [see figure 5.5, *top*], and if somebody says they have a temperature, my mind goes to the area above 100°—to that corner." As she ages, the forms that relate to her personal age grow from age 0 like a vine.

Like Marti and Colleen, Tammy Lorch experiences forms for concepts involving sequences beyond the typical centuries, months, weekdays, and integers. For example, her form for salaries starts at $1,000 on the right, with higher figures moving leftward at different angles. TV channel 1 is on the right, and higher channels move diagonally to the left. For radio stations, lower channels start on the right and higher channels move diagonally right. She also has patterns for shoe and clothing sizes that include half sizes.

Another spatial sequence synesthete, Deborah R, does not so much sense a spatial configuration as she does a spatial location to which she *goes* or *reaches*. "I'm not sure I really see anything, but I go to a place where the number 1 or whatever is." Simple line drawings fail to convey the panorama and depth she senses. "[My] drawings aren't accurate because in my mind they are three-dimensional: either lying flat *in a single plane* or else *coming at me* at an angle and *crisscrossing* through other horizontal planes."

Visualized timelines are sometimes set against a definite background. Marti Pike's forms are all set against black, Nancy T's number line appears against a backdrop of starry space, as in a frame from *Star Wars*, whereas Marcia Smilack sees decades against a background of a geographical map:

A decade is a three-dimensional rectangle in bluish-gray like the uniforms worn in the Civil War. It is shallow ... [and] suspended in space in front of me slightly to my left in a horizontal position. ... It is situated right in front of Chicago. I know that sounds funny, but it is true and this is how it works. The spatial plane on which a decade is suspended is a transparent plane; that is why I can see geography behind it.

What appears in the background depends on what timelines Marcia attends to: she sees background geography maps when she thinks about decades, but not centuries or the months.

Many synesthetes we have worked with have generously spent time painstakingly drawing their various forms to help us better understand what they experience. Until now, however, there has been no systematic way to study the similarities, differences, patterns, and changes in number forms in real three-dimensional coordinates. To address this problem, David's laboratory has developed virtual reality software that allows synesthetes to physically place their time units exactly where they perceive them in three-dimensional space. Figure 5.6 illustrates the virtual reality space of two sisters' number forms. Note that even though the sisters were raised in the same household, there is almost no similarity between their forms. This speaks against the possibility that number forms are something that children learn from their parents or from exposure to some strange calendar in their home.

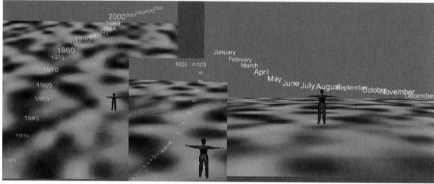

Figure 5.6
Virtual reality placement of number forms by two sisters. The one pictured on the top has forms for months, weekdays and numbers, while the one at bottom has forms for years, numbers, and months.

Using virtual reality methods to determine exact three-dimensional co-ordinates opens the door to two possibilities: finding patterns across the spatial coordinates of different subjects and understanding how forms change over time. Subjects asked to mark their places in three-dimensional virtual reality space revisit the laboratory a year later to place their sequences again. This is done without warning, and the test–retest results are compared. This work in progress seems to indicate that the three-dimensional coordinates for weeks, months, and number lines change little over time. On the other hand, forms for age and years show greater change because the contexts of those sequences vary as the subject ages. In one sense, the subject is "standing" in a different place in history from one year to the next, while Monday through Sunday remain immutable as time passes.

Among its advantages, this method of quantifying three-dimensional coordinates can rule out malingerers. Nonsynesthetic subjects asked to use their imagination in placing their sequences and tested twice over the period of one month showed little, if any, correlation among their forms.

Common Motifs?

Do various number forms share common features? Most obviously, they all involve learned sequences, suggesting the possibility of a specific window during mid-childhood when the gene expresses itself, a prospect we explore in chapter 9.

For now, we can note some motifs common to different number forms. As with color → grapheme synesthesia, there may be a bit of imprinting in individuals with number forms, meaning that they can be influenced by exposure to a pattern seen in everyday life. Consider, for example, the often-seen placement of numerals 1 to 12 in a clocklike pattern. Figure 5.7 illustrates two number lines: one from a modern day subject in Paris,[3] the other from a subject reported by Sir Francis Galton in 1883. Both contain the clock face motif, as does Marti Pike's number form for the integers (see

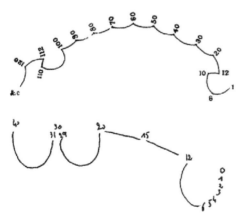

Figure 5.7
Commonalities among number forms. (*Top*) Number line reported by Sir Francis Galton in 1883. Note the clocklike pattern of numerals 1 to 12 and also the scalloping between the landmark decade numbers, themes commonly expressed in number forms. (*Bottom*) Integers of a subject in Paris (from Piazza et al., 2006).

figure 2.5). Note that Marti's "clock" is upside down and going in the wrong direction, something she had to unlearn when learning to tell time. Thus, she seems to have imprinted on clock faces during the childhood window during which she was learning numbers, but before understanding what the clock face arrangement *meant*. This puts the likely window for the development of number forms at an early age, probably before forty-eight months.

Another motif commonly observed is the recurrence of landmark numbers such as the multiples of 10 in figure 5.6 or in Marti Pike's decades illustrated in figure 5.8. Grouping by 10s appears to provide structure for the in-between numbers. This motif is not surprising, given the use and utility of landmark numbers in nonsynesthetes: they may be a key to how nonsynesthetes commonly do mathematics in their head,[4] an issue to which we return below.

Note in figure 5.8 that the decades of the 1900s rest upon the blue baseline of their century. "Thinking of each century as a whole," says Marti, "1900s are blue, 1800s—yellow, 1600—red, etc." However, when she zooms

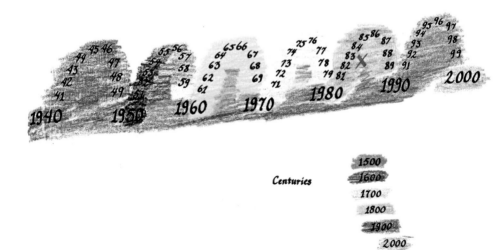

Figure 5.8
Marti Pike's decades are organized in groupings of 10—a "landmark" number. Note its similarity to her integer form and its colors in figure 2.5.

in, "the decade color takes over." The shape of her form is obviously similar to that for her integers depicted in figure 2.5, although there are exceptions in coloration (e.g., she has different colors for 50 and 1950, and 80 and 1980).

Another common motif involves the past, present, and future. In interviewing many individuals with number forms, it is striking that the future, which cannot be seen, nonetheless has a spatial representation. For many, the future lies either behind the person or so far distant to the front that they cannot clearly discern it. The same applies to the distant past: it is often in a location difficult to make out, either because it is far behind the subject or too small to resolve. Observe Marcia Smilack's report:

Each era is a thick slab though I only see the face of the nineteenth century; for the others, I "know" they are arranged in a descending chronological order that *reaches back beyond where I can see*; after around the fourteenth century, they *become so small they are hard to identify* and *further back than that, they disappear* from view . . . [*italics added*].

The present is often seen from a viewpoint of where individuals are spatially standing now. It is more physically discernable than either past and future. For example, the years long before Marti Pike's birth are dark. "When I was perhaps 10 or 11, I asked mom if she grew up in the 'Dark Ages.' She laughed, of course, having no idea what I was thinking, *But at the time*, anything before 1954, my birth, *was dark in my mind* [*italics added*]." As Marti matured, the decades and centuries before her birth decade became both colored and shaped (the pre-1950s remaining gray and dark, however).

Negative numbers and B.C. years are similarly "in shadow, and have very little, if any, color." Negative numbers are seen "below zero," and B.C. dates are "mirror images to the left of zero" (see figure 5.9).

In this way, the timelines of years tend to be different from the timelines for months and weekdays, which almost always lie on circular, elliptical, or horseshoe paths within easy reach—see, for example, figure 5.6 (*top*).

Aside from clock faces, scalloping between decades, and the spatial proximity of the present together with the distance of past and future, it is not clear that any consistent pattern exists as to the direction in which number forms extend. It is rarely the case that time-spans simply travel from left to right, or that January or Sunday are always at the top. The only feature

Negative numbers, Years BC, etc.

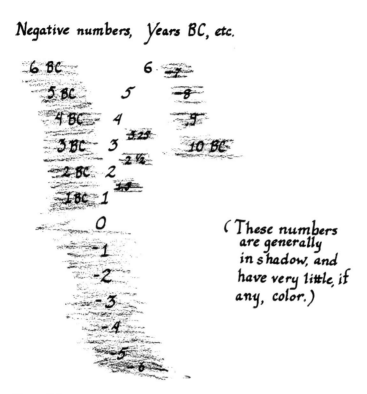

(These numbers
are generally
in shadow, and
have very little, if
any, color.)

Figure 5.9
Marti Pike's colorless negative numbers and B.C. dates lie below zero.

that seems consistent is that appropriate relationships are maintained between neighboring elements in a sequence. In other words, a synesthete could, in theory, experience Wednesday next to Saturday next to February, but instead sequential elements always end up as neighbors in space. This simple observation serves as a clue as neuroscience plumbs deeper into the origins of number forms.

Finally, we have often emphasized in previous chapters that synesthesia becomes fixed once associations are established at an early age. Number forms, however, may be an exception to that rule: they sometimes change or "grow" with age. We mentioned Colleen Silva above, who remarked that number forms relating to her age grew from their terminal end "like a vine." Figure 5.10 illustrates how she perceives her age and history in general and how the forms have changed over time. Note that as she matures from a teenager to a young adult the pattern not only extends but

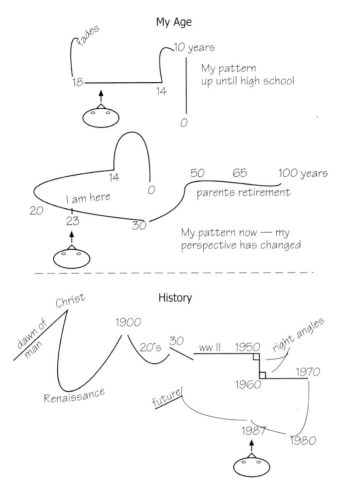

Figure 5.10
Colleen Silva's perceptions of her age and history. Her personal forms have changed over time.

changes its shape. Clearly, new learning is going on, but the obvious question is "What is it about number forms that lets experience continue to modify them?" This question remains open for future study.

A Surprisingly Common Phenomenon

When David began advertising for subjects with synesthesia, he expected the prevalence to be around 1 in 2,000, as Simon Baron-Cohen and colleagues had calculated in their 1996 survey. To his surprise, however, he found synesthesia to be much more common. We now know from Julia Simner's random population survey that the chances of having any type of synesthesia is as high as 1 in 23. A surprise to David was that about 25% of the synesthetes responding to his advertisements around the Texas Medical Center had spatial sequence synesthesia. This apparently rare form of synesthesia did not seem to be so rare anymore—in fact, it was beginning to look like the most common known form.

In 2004, Noam Sagiv and colleagues[5] set out to study exactly how prevalent number forms are in the general population. To get a large and relatively random sample of people, they set up tables in the London Science Museum and recruited volunteers who were passing through the museum that day. Out of 1,000 naive volunteers, they discovered that number forms were quite common, perhaps as common as 1 in 10 people. This finding was roughly consistent with earlier estimates (from nonrandom samples) of 4.5% to 12% of the population.[6] Sagiv's findings further suggested that number form synesthesia is highly correlated with other varieties: if you have taste synesthesia, for example, the chances of experiencing number forms increases to 20%; if you are a grapheme → color synesthete, the chances escalate to 60%. These observations shed light on the neurobiological basis of synesthesia, which we explore in chapter 9.

Prevalence estimates have to be taken with a bit of caution because it is somewhat difficult to nail down who genuinely has spatial sequence synesthesia and who does not. To illustrate the problem, ask a dozen friends if they experience number forms. A few of them will know exactly what you are talking about, whereas most will think you have just come from another planet. The difficulty is that there are intermediate cases wherein a friend might assert that they "think of" the weekdays running straight from left to right they way one sees them printed on a calendar. Is this synesthesia, or is it merely the way they think the question should be answered if it were to have an answer at all? What this question—and synesthesia in general—highlights is that different people think in widely different ways. Some people are highly visual, whereas others are not. As we indicated earlier, synesthesia may lie on a continuum with ordinary perception.

Researchers are beginning to develop tests for measuring behavioral consequences of number forms.[7] For example, are spatial sequence synesthetes better at mathematics given that low visualizers are worse than high visualizers at math skills?[8] We do not yet know the answer, but it is a testable possibility. Another possibility is that measurable reaction time differences exist depending on the exact shape of an individual's number line. That is, measurable differences might be detectable when synesthetes are asked to compare numbers that lie close together compared to far apart on their contorted lines.

The reason to suspect this might be true is the evidence that everyone, synesthete and nonsynesthete alike, makes automatic associations between numbers and space. A straightforward demonstration of this is an effect discovered by Stanislas Dehaene and colleagues called Spatial Numerical Association of Response Codes, or the SNARC effect.[9] Imagine being asked to indicate whether a one-digit number flashed on a computer screen is odd or even. For some trials, you respond with your right hand to indicate "even." For other trials, you use your left hand. It turns out you respond more quickly to small numbers (0–4) with your left hand, while you respond to larger numbers (5–9) more quickly with your right hand. This finding, replicated many times since, demonstrates that numbers are automatically associated with positions in space, supporting the idea of a spatially organized internal "number line" in everyone, synesthetes and nonsynesthetes alike. Moreover, when subjects are asked to cross their hands, their responses cross too—being faster to a small number with the right hand now on the left half of space—indicating that it is not the hand but the side of space that matters. A similar finding occurs with eye movements: subjects look faster to the left when responding to small numbers and to the right when numbers are large. The direction of the effect is sensitive to cultural experience: Iranian subjects, who write from right to left, show the SNARC effect in the reverse direction, whereas Japanese subjects, who write both right to left and top to bottom show it in both left–small and bottom–small directions.

In another demonstration of the implicit spatial organization of numbers, subjects are asked whether a number is smaller or larger than 65. Individuals respond faster to smaller numbers with their left hand, and faster to larger numbers with their right one. Interestingly, responses are more rapid with increasing numerical distance from the number 65—so individuals are faster when indicating that 97 is larger than 65 than they are when responding that 67 is larger. The difference is thought to reflect a larger "distance" on their mental number line.

Observing that synesthetes have spatial forms for the same kind of sequences for which snarc effects exist in nonsynesthetes (namely, number lines), Ed Hubbard and colleagues have begun to make explicit connections between spatial sequence synesthesia and the SNARC effect.[10] One similarity between synesthetes and nonsynesthetes, for example, is that the mental number line compresses as it reaches successively larger numbers. In other words, your brain devotes more resolution to 13, 14, and 15 than it does to 10,713, 10,714, and 10,715. This compression is evident in the number forms shown in figures 2.2, 2.6, 5.6, and 5.10. The similarities between nonsynesthetic and synesthetic number representation has caused a renewed examination of how numbers are represented in the brain. (There are at least three ways: verbally, visually, and quantitatively abstract, all of which are associated with different brain structures.)[11] It has also led to appreciating the way in which the neural representation of number is inextricably related to the neural representation of space.[12] Given the SNARC effect in nonsynesthetes (which reveals an implicit mental number line) and the number forms of spatial sequence synesthetes (who explicitly experience mental number lines), it is clear that all of us represent numbers spatially in our brains while only a subset have vividly conscious access to these associations in three dimensions.

Sequences and the Philosophy of Time

The spatial relationship among numbers in a form is reminiscent of a philosophical point of view of time that has increasingly gained acceptance. In Kurt Vonnegut's *Slaughterhouse Five*, the protagonist, Billy Pilgrim, fresh from World War II in Europe, is kidnapped by aliens called Tralfamadorians and put on display in their zoo. They are bemused by the Earthling's notion of time in which there is a present "now" with everything else belonging to a future or a past, because for Tralfamadorians time exists all at once. For them time is like looking at the Rocky Mountain range: you see the whole thing in one view. You can pay attention to one part or another, but the whole vista is always there, always available to be seen.

The Tralfamadorians have what philosophers call a "tenseless" way of looking at time, in which past, present, and future exist equally.[13] A similar point of view occurs in H. G. Wells's *The Time Traveler*. In the tenseless view, past, present, and future are not the important properties to worry about: it is only the *relationship* between events that matters, and the possible relationships are "before," "after," or "simultaneous with." Past and future are analogous to left and right—that is, they are relational.[14] In this view, there is nothing special about the present. It merely holds relationships to things to the right or left of it. It is merely one peak on the Tralfamadorians' Rocky Mountains. Thus, both to the philosopher who holds the tenseless view of time and, perhaps, to the synesthete, your birth lies to the left of your picking up this book, which is itself to the left of your death. The experience of the spatial sequence synesthete gives physical form to this view of time.

Conclusions

Number forms have been commented upon for over a century, but only recently has it been appreciated how common they are. An estimated 10% of the population experiences some form of sequence as having spatial dimensions. It is becoming increasingly clear that the representation of number and the representation of space share the same neural areas and that damage to these areas typically leads to deficits in both space and sequences. Strange numerical effects in nonsynesthetes (such as the SNARC effect) indicate that numerical sequences are represented in spatial layouts in all brains. For 10% of the population, this spatial layout becomes explicit (we will explore the reasons for this in chapter 9). For that segment of the population, sequences and space are overtly linked. As Marcia Smilack points out, "One of my constant adjustments as a synesthete is to remember that for other people, time and space are distinct and different."

6 A Matter of Taste

Taste and smell are intimately related—so much so that people are surprised to learn that what they think of as taste is more often a matter of smell. Foods lose their flavor when you have a cold because of your diminished sense of smell. What people commonly experience as flavor is actually a *compound sensation* of discriminating basic tastes (sweet, salty, bitter, sour, meaty)[1] along with smell, temperature, and texture. Compared to our small handful of taste receptors, we possess about 1,000 olfactory receptors, the result being that aroma sensations are far more diverse than taste sensations.

To prove the importance of smell to tasting, you have only to hold your nose or put a clothespin on it while sampling a variety of foodstuffs. Do this and you will discover how bland or tasteless many foods become. For example, you will be unable to distinguish an apple from an onion, or coffee from tea. The latter two will taste only mildly bitter and unpleasant because it is their robust aromas that give each beverage its characteristically complex flavor. In using the term "flavor," we therefore mean to include both taste and smell unless specifically stating otherwise.

Taste and smell differ from other senses in being chemically based—that is, they rely on the dissolving of chemical molecules in the tongue's taste buds or mucous membranes of the nose and throat, as opposed to the transduction of electromagnetic forces (vision) or mechanical ones (touch and hearing). As such, taste and smell are organized much differently in the brain than vision, hearing, and touch are. For one thing, the nerves carrying smell information do not pass through the waystation of the brain's thalamus, instead synapsing directly in the cortex at the bottom surface of the frontal lobes. This ancestrally old part of the brain is called the "rhinencephalon," literally the "smell brain."

The primary cortical taste area lies in a spot on the side of the brain called the operculum, with secondary taste areas further back in the parietal lobe and deeper in from the temple within a region called the limen insula (see figure 6.1). Damage here impairs the recognition of food type and flavor intensity.[2] Figure 6.2 shows the cortical components of an aroma network.[3] Over half the neurons in the primary taste areas are also sensitive to texture and temperature.

It is little appreciated how cross-sensory perceptions occur routinely in olfaction. In a technical sense, we might say that everyone is synesthetic.[4] For example, certain odors such as vanilla are consistently said to smell sweet, even though sweetness belongs to the domain of taste.[5] In fact, "sweet" is the most common description of odor. When one researcher asked 140 participants to describe a strawberry odor, 79% said it smelled sweet whereas only 43% said it smelled like strawberry and 71% that it was fruity.[6] Many similar examples (such as amyl acetate, commonly used as banana flavoring) yield the same result: when smelling an odor, more

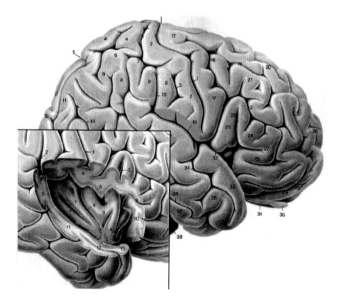

Figure 6.1
Primary cortical taste areas, in green, are in the fronto-parietal operculum, and, in the cutaway, the insula. There are additionally many nuclei in the brainstem and amygdala that also participate in taste as well as smell.

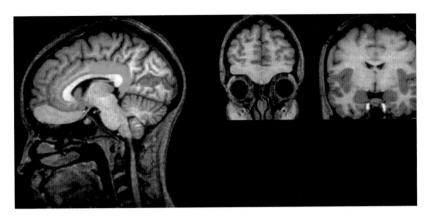

Figure 6.2
The primary smell cortex lies on the bottom side of the frontal lobe (yellow). An extended aroma network includes the subcallosal gyrus (blue), amygdala (red), visual area V1 (orange), hypothalamus (green), and temporal pole (not pictured). The insula, outlined in purple, at its most forward part relates to smell–taste–autonomic function.

people perceive taste-like qualities such as sweetness rather than specific ones such as strawberry-likeness or banana-likeness.

Given the dramatic effect of smell on taste perception, it is therefore ironic that we have few, if any, words to primarily describe smells. Rather, we borrow nearly all of them from other senses as revealed by terms such as "sweet," "sharp," "bright," "clean," "fresh," "soft," or "spicy." Terms used for smells usually refer to their typical causes, for example, "floral," "fruity," "moldy," or "acrid."

As in true synesthesia, the reliability of ratings such as sweetness as measured by test–retest correlations is stable over long periods of time. Their validity is likewise shown by the phenomenon of sweetness enhancement. That is, when a sweet-smelling odor is added to a sugar solution, its sweet taste is enhanced (a phenomenon exploited by commercial manufacturers). The reverse of enhancement is also observed wherein sweet-smelling odors reduce the perceived sourness of acidic solutions.

Additionally, the taste properties of an odor are affected by a history of simultaneously occurring with sweet, sour, bitter, or salty tastes in the past. Odor–taste learning is largely implicit, meaning it occurs along unconscious lines without the need for deliberate intention.[7]

Straightforward experiments with taste and smell indicate that how we attribute perception to a given sense domain may be malleable even in adults. For example, there is now direct evidence[8] that odor qualities can be acquired. For odor mixtures such as mushroom–cherry, the subsequent smell of mushroom alone takes on a cherry aroma. A surprising feature of such an odor memory is its resistance to interference by later experience. Some researchers therefore argue that the commonly occurring taste properties of odors should be regarded as a form of synesthesia.

Flavor Triggers

Taste and smell can trigger synesthesia as well as being a synesthetic experience. We will consider the triggering sensations first and the person who started the modern renaissance of interest in synesthesia back in 1980, Michael Watson. His is probably the best-known instance of flavor triggering touch sensations.

What Michael senses are shapes, textures, temperature, and weight, all various aspects of touch. His parallel perceptions are especially sensuous and vivid when he encounters novel flavors, meaning those sampled for the first time. Naturally, he likes going to restaurants:

Eating for me is an impulse, and the first bite I take of a new course is an urge for me to look in a new direction; I feel drawn by this. These new experiences are frequently very erotic, in the term of sensual, although sometimes they are erogenous as well . . .[9]

His perceived sexiness of food should not surprise us given how tightly flavor perception is bound with the limbic brain, which affects libido (think also of the intriguing evolution of the kiss). On the positive side, neuropsychologically, Michael also has eidetic memory, while on the negative he has difficulty with math and confuses right with left.

Michael has an excellent memory for configurations, effortlessly remembering shapes he has encountered before but not the flavors that go with them (the latter a result of his synesthesia being unidirectional). Michael enjoys cooking but never follows a recipe. Instead he guides himself by a rough idea of what he wants the final dish to feel like, adjusting the ingredients and seasonings by trial and error—for example, altering the flavor's shape to make it "rounder," giving it more "inclination," "sharpening up the corners" to give the vertical lines more heft, or adding "a couple of points" to the overall shape. His groping method is simply an example of knowing what you are looking for once you find it; when a dish's shape eventually matches his starting idea, a "eureka" feeling comes over him.

Michael, who passed away in 1992, would have loved the Web site of "synesthetic recipes" created in 2005 by MIT's Media Lab[10] despite the fact that it contains no shape terms. Housing 60,000 recipes, its graphical interface allows cooks to brainstorm dinner ideas by describing how they want to dish to feel (e.g., "spicy," "chewy," "moist," "fluffy," "rich"). The search agent accepts 5,000 ingredient keywords (e.g., "chicken," "blueberries," "cinnamon"), 1,000 taste-space words (e.g., "spicy, "mushy"), 400 nutrient words (e.g., "beta carotene," "zinc"), and all negations (e.g., "no beef," "not spicy"). Taste preferences of family members can also be input to please every taste.

Having discovered at an early age that others do not taste shapes, Michael habitually kept his synesthetic percepts to himself to avoid ridicule and disbelief. Nonetheless the sensations were automatic and impossible to ignore. Not surprisingly, he was polymodal: a flavor sometimes caused him to "see" a colored pattern. For example, orange extract caused "olive green patches through a rectangular doorway." The perception was vivid enough to block his central vision so that he could not see through the patches. More often, sounds had shape:

When I was in the third and fourth grade there as a radio program called "Musical Pictures" the class would tune to every Thursday. We would draw what the music said, and I was fascinated by it. I could draw things, I was wonderful. I was the best person in the class, just being able to visualize things from sound. I remember it vividly and it was my favorite part of those years.[11]

Michael felt tactile sensations "throughout the body," but mostly in the face, hands, and shoulders. The sensation was strong and pleasurable, and felt "as a rush." Often he had a sense of grasping or manipulating an object, of fingering its texture and temperature with both hands. At times his thumb or middle finger felt something more intensely than his other digits. The homunculus in figure 6.3 suggests why these body parts are predominantly affected given that they take up more cortical area than

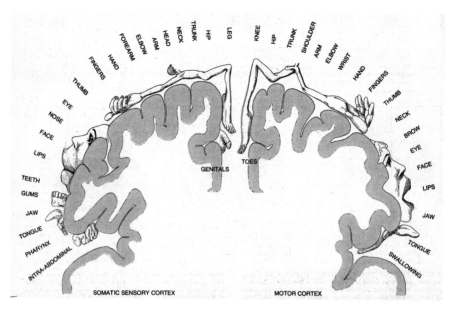

Figure 6.3

Most of the body can be mapped onto sensory (*left*) and motor (*right*) cortices. The distorted homunculi result because the area of dedicated cortex is determined not by the size of the body part it innervates but by the precision with which it must be controlled. This is why the sensory and motor regions dedicated to the face and hands are exaggerated compared to the rest of the body. Adapted from Penfield and Rasmussen (1950).

the trunk or a leg does. Movement was nearly always present—Michael said, "A sensation sweeps down my arm into my hand. The thumb runs along the edge to determine the boundaries while the fingers feel the texture." The taste itself also moved, starting in the mouth but then "rising up" into Michael's head. Almost always his perceptions were pleasant, rarely a "slap" or "burning" in his face in response to a "sharp" or "pointed" shape such as lemon or vinegar. When he felt a particularly intense or satisfying synesthesia, Michael could not help but pause the conversation, gesture, or engage in some other body language that signaled to others an attentive absorption in his private sensuous world.

Michael wondered why only certain fingers rather than his whole hand felt the taste sensations. The same mystery applied to his shoulders and arms. Common sense told him that the feeling should be equally spread out, but instead he felt it in long bands traveling from his shoulders to his

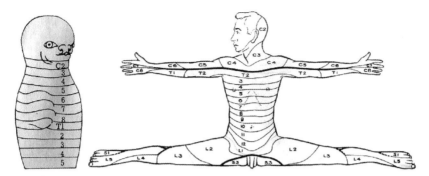

Figure 6.4
Dermatome chart. (*Left*) The serial segmentation of the embryonic body is disrupted by the development of the head and limbs, causing skin map (dermatomal) disloca-tion (*right*) such that nonsequential areas of innervation abut in the adult. Michael Watson variously felt tactile taste sensations in dermatomes C2 through T4.

fingers. When shown a dermatome chart illustrating how the extremities are actually innervated, he was relieved, "So that's why I only feel it in certain parts!" (figure 6.4).

His descriptions were highly detailed. Banana extract was "round and carved, like Baroque molding," camphor "like a rectangular handle on a briefcase," honey "long and linear with bumps, like a polished walking stick." Smelling peach preserves made him feel a sphere, but tasting it added holes "like a bowling ball" so he could grab onto and sink his fingers into it. Menthol felt "tantalizing and odd," compelling him to turn his head to the left and "move around" the shape. "Something is around the corner, leading me forward," he said. Strawberry extract was not only the rounded "top half of a sphere" but also intense and "sexy—a 10 on a scale of 1 to 10." He felt it in his face and neck all the way down to mid-chest level, "the farthest down I've ever tasted anything."

In trying to characterize which kinds of flavors produced which kinds of shapes, Richard first compiled the most common adjectives Michael used to describe them: "linear, pointed, globular, undulating, planar, rough, earthy, polyhedral, tubular, spherical, angular, smooth, polished, cold, slapping, and prickly." He then constructed a circular response set of shapes for Michael that ranged from completely round (spherical) to com-pletely angular (cubic), as shown in the left half of figure 6.5. Pilot experi-ments using ten varied flavors—salt, sucrose, anise, citric acid, Campari,

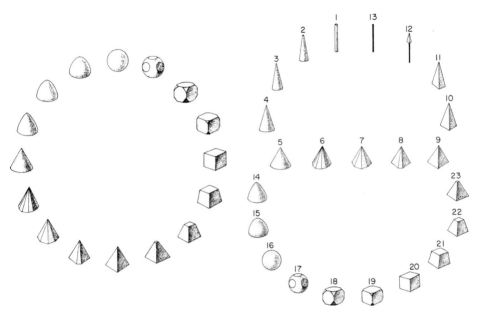

Figure 6.5

Array of shapes from which Michael Watson could choose in taste–shape matching experiments. (*Left*) Pilot experiments showed an orderly progression from completely spherical to cubic to be an inadequate choice doman; (*right*) revised circumplex array.

menthol, Angostura bitters, vanilla, quinine, and Karo syrup—revealed that symmetrically radial shapes were inadequate given that Michael often reported sensations that were linear, columnar, or pointed. Consequently, a revised figure-eight response set was devised (figure 6.5, *right*) to incorporate a wider array of shapes that differed from one another along more than one orderly series of intermediate configurations.

Michael then underwent psychophysical experiments, meaning the systematic mapping of stimuli to responses to see what patterns, if any, emerged. Patterns can usefully guide ideas of how synesthesia is represented in the nervous system. Michael tasted 13 solutions of equal concentration that ranged from completely sweet (sucrose) for solution No. 1, to completely sour (citric acid) for solution No. 13. Solution No. 7 was a 50:50 mixture of both. The different proportions of sweet to sour enabled the testing of three different flavor ranges: a sweet to mildly sour range (Nos. 1 to 7), a mildly sweet to fully sour range (Nos. 7 to 13), and an

Table 6.1
"Low," "High," and "Extended" arrangement of taste stimuli

Sweet						Sour						
Experiment I												
1	2	3	4	5	6	7						
Experiment II												
						7	8	9	10	11	12	13
Experiment III												
1		3		5		7		9		11		13

(From Cytowic and Wood [1982, p. 41], with permission.)
Note. See text for explanation of experiment numbers.

extended range from wholly sweet to wholly sour made up of the seven odd-numbered solutions (see table 6.1). It is common knowledge that ordinary people show predominantly contextual effects in perception. That is, they judge the 50:50 mixture as sour when it is at the top of a flavor range and sweet when it is at the bottom. Did Michael perceive tastes absolutely regardless of their context, or did a flavor's relative sweetness or tartness determine the shapes he picked? The experiment set out to disprove the hypothesis that Michael would be no different from nonsynesthetic controls in matching tastes to shapes.

To make it impossible for Michael to keep track of his answers he tasted 147 separate samples of the 13 solutions over multiple trials. He performed markedly unlike controls. Whereas controls spread their choices out over available selections, Michael's responses clustered absolutely in one part of the response domain (the acidic half). In this range he used essentially only three shapes, all some pointed, angular shape conceptually close to one another. By contrast, in the sweet half Michael's choices were just as spread out as when he responded to the entire spectrum of solutions. Thus, he was sensitive to context (how relatively sweet or sour a taste was) in one part of the flavor domain, while making context-free matches in the other part. The likelihood of his doing so by chance was nil.[12]

The mixed result indicates that in terms of the nervous system level at which synesthetic links are established, the stimulus–response association is neither absolutely hardwired one-to-one nor contextually one-to-many. Rather, the combination of absolute and contextual effects supports the notion of an intermediate middle ground of stimulus–response mapping.

A similar setup of testing "high," "low," and "extended" ranges of a syn-esthete with colored hearing (VE) yielded similar results.[13]

A Case of Shaped Smell

The British psychologist John Harrison describes the case of AJ, who is similar to Michael Watson except for having her synesthetic shapes trig-gered by smell rather than taste.[14] For stimuli she was given the standard-ized Smell Identification Test that had been invented at the University of Pennsylvania by Richard Doty a few years after Richard encountered Michael Watson.[15] The test consists of forty scratch-and-sniff patches, which makes test–retest sessions easy. AJ was given a so-called forced-choice paradigm wherein she had to match smells to one of only four possible shapes. Table 6.2 shows her results.

The small number of permitted choices is unfortunate as is the manda-tory use of verbal labels to describe shape instead of letting her freely draw, as had been encouraged in Michael Watson's case. Nonetheless, all four form constants are recognizable even in her restricted descriptions—grids, spirals ("drill bit"), cones ("taper"), and central radiations ("dish," "round shape," "rising dough").

AJ underwent neuroimaging. Her activation task was breathing a sequence of three odors (clove, wintergreen, menthol) for thirty seconds, whereas her control task was breathing odorless air. Despite feeling shapes during the activation task, AJ's brain maps showed nothing different from those of nonsynesthetic controls. The failure to find unique foci of brain activation or deactivation is perhaps due to the experimental design, the technical resolution of scanners at the time, and/or the small sample size.

Michael Watson underwent a technologically earlier form of brain mapping called xenon-133 regional cerebral blood flow (rCBF) that had dramatically less resolution than today's brain scanners do. While it did not localize his synesthesia to any particular area, it did confirm that Michael's brain was unusually reactive to smells, especially in the left hemisphere.[16] Of greatest note were widespread and profound *deactivations* (reductions in activity) in the motor and sensory areas of the face and hands when presented with a smell. This is precisely the opposite of what one would have expected during his synesthesia. Further neuroimaging

Table 6.2

Shapes smelled by AJ

Odor	AJ's identification	Percept description
pizza	pizza	black flex arrow from top
bubblegum	bubblegum	wide, all filling
menthol	menthol	tall shape, not quite a column, curls a bit at the top
cherry	cherry	wave shape
motor oil	motor oil	mushroom
mint	mint	flat, but not filling like bubblegum
banana	banana	round shape
cloves	cloves	spearhead shape
leather	leather	lip at bottom
coconut	coconut	spread out shape
onion	onion	collection of grids
fruit punch	fruit punch	mushroom spiraling under the cap
gingerbread	gingerbread	arrows, looking down on points, prickly
lilac	lilac	shaped like a drill bit
peach	peach	wide smell that tapers at the top
root beer	root beer	thin, high, rising shape
pineapple	pineapple	layers of smell together
lime	lime	flat with edges, smooth like
orange	orange	black bits, a tall smell, about 2 foot
wintergreen	wintergreen	ragged edges
watermelon	watermelon	flat dish shape, could be circle of spears
grass	grass	flat, wide smell
smoke	smoke	spearhead shape
pine	pine	upward moving
grape	grape	big and filling like rising dough

with more modern technologies will be needed to verify and understand these results.

Drug Effects on Synesthesia

A second activation state was used during Michael's study. It involved the inhaled drug amyl nitrite, behaviorally known to intensify his synesthesia. Although it did not help sort out the unusual metabolic landscape of Michael's working brain, it did suggest that different classes of drugs could either intensify or weaken synesthetic experience. Serendipitously, Michael's pattern of feeling synesthesia less in the morning compared to the evening, together with his dietary history, suggested two candidates: caffeine (a nervous stimulant) and alcohol (a nervous depressant). Michael's standard breakfast consisted of cigarettes and lots of coffee, a doubly stimulating combination because nicotine is also a nervous stimulant. Dextroamphet-amine was substituted as the stimulant in the drug experiments because its dosage could be standardized.

The synesthetic stimulus was spearmint, which reliably caused Michael to feel cold, smooth glass columns along whose curvature he could run his hand. When he was under the influence of dextroamphetamine, the columns became "much smaller . . . more distant . . . like a scale model or miniature." When the amphetamine had its peak pharmacological effect, his synesthesia was completely blocked.

Amyl nitrite normally reduces cerebral blood flow by 5% to 10% (a fact confirmed in Michael's neuroimaging study).[17] Its effect on synesthesia was to enhance it markedly, by making the touch sensations Michael felt more vivid, textured, and numerous without changing the basic shapes he nor-mally felt in the absence of amyl nitrite. One and a half ounces of absolute ethanol (200 proof) also increased the magnitude and intensity of his synesthesia during the time it exerted measurable pharmacological effects. Table 6.3 summarizes the effects of the three drugs on the brain and on Michael's synesthesia. In general, this led Richard to hypothesize that corti-cal stimulants will tend to suppress or block synesthesia, while cortical depressants will tend to intensify it.[18]

A natural experiment then occurred which supported this hypothesis. When his synesthesia first came to Cytowic's attention, Michael's average alcohol consumption was eight ounces a day. That consumption escalated in 1983 (two years after the experiments mentioned here) until he was consuming nearly a fifth of liquor a day at the time he stopped drinking

Table 6.3
Drug effects on synesthesia

Drug	Cortical effect	Effect on synesthesia
Dextroamphetamine	Stimulates	Blocks
Ethanol	Depresses	Enhances
Amyl nitrite	Ischemia	Enhances

in April 1985. During these years he enjoyed his synesthesia immensely, but following the total cessation of alcohol his synesthesia was markedly diminished for several months. He thought it had simply "faded away" and mourned the loss, whereas in fact the results were predictable. Pharmacologically, he had been chronically exposed to a cortical suppressant; its sudden removal resulted in a well-known opposite effect of rebound hypersensitivity and overactivity—in effect the same as if he had been given a stimulant.

Based on the drug effects in Michael Watson, David surveyed prescription and social drug use in 1,279 verified grapheme → color synesthetes, looking for suggestive patterns (see figure 6.6). Although this can only count as an informal survey, it seems clear that the majority of those who volunteer reports find their synesthesia to be *unchanged* by alcohol, cigarettes, and caffeine. The category of "drugs" remains difficult to interpret, since at this stage we did not probe for details about the type of drug used. The figure illustrates two points: first, synesthesia can wax and wane in strength—it appears that under conditions of fatigue and high emotion, many report a modulation of the synesthetic experience. However, common nervous system stimulants and depressants appear to have little effect on most synesthetes. Moreover, those that do report an effect may be

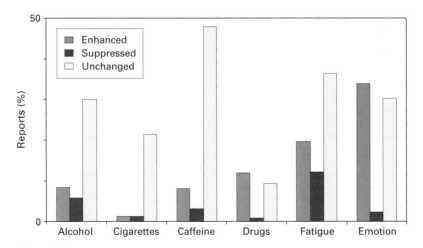

Figure 6.6
Answers to the question "How does _____ change your synesthesia?" Gathered from 1,279 verified grapheme → color synesthetes on www.synesthete.org.

conflating indirect effects of fatigue or emotion associated with the drug use. Therefore we cannot say that stimulants and depressants will *inevitably* influence synesthesia, although it remains a possibility that some drugs with specific neural targets (such as dextroamphetamine, LSD, anti-epileptics, and others) are more likely to produce clearly measurable changes. This is a subject for future study. In chapter 9 we suggest some specific candidates for molecules that may be effective in modulating synesthesia, examining how changes in inhibition can allow one brain area to causally affect another.

Colored Tastes and Smells

More common than shaped tastes are colored tastes and smells. Given the intimate relationship between the two senses, it is peculiar that some individuals insist only one sense and not the other triggers their synesthesia. For example, Muriel Nolan is a polymodal synesthete for whom sound → touch, color, location. However, she experiences only smell, and not taste, as colored:

I remember most accurately scents. We were preparing to move into the house I grew up in. I remember at age 2 my father was on a ladder painting the left side of the wall. I remember to this day thinking why the paint was white, when it smelled blue [see figure 6.7].

Sean Day is especially fond of foods that taste blue, such as milk, cheese, all citrus fruits, vanilla, beef (dark blue), buffalo (darker), and chicken (light blue).[19] A concoction that tastes a particularly wonderful shade of blue consists of pasta with tomatoes, ricotta cheese, nutmeg, and spinach, whereas his "chicken à la mode with orange sauce" combines different textures with different shades of blue. It consists of baked chicken with a scoop of vanilla ice cream topped with orange juice concentrate. "It's quite delicious," insists Sean. Most of what he sees he projects spatially, at eye level and about a yard away.

Downey reported the earliest case of colored taste in 1911,[20] her subject experiencing color only for taste and not smell. His colors were spatially extended, being felt in different locations in the mouth and sometimes projected onto external objects. In general, pink and lavender tastes were

Figure 6.7
A colored smell—and an example of sensory conflict of white paint smelling "blue."

agreeable; reds and browns were not (and for some reason, "blue tastes are never experienced"). When actual food colors conflicted with synesthetic colors, the result was "most disagreeable." Sweet tastes were black, sometimes "brilliant," bitter was dull orange–red and burning, salty tastes were crystal clear, and sour tastes were green and cool. Downey's subject was also polymodal, ascribing color to touch. Green had "an agreeable feel" but was not an agreeable color. Blue–greens caused a "perfectly awful feeling, like . . . sandpaper; disagreeable to both sight and taste." The color of lime candy was "pretty," while its taste was not "particularly agreeable."

The natural color of foods appears to play no part in determining taste colors. As Rebecca Price puts it:

My mother says I have always used colors to describe flavors. Kiwi fruit tastes green, and there is no other way to describe it. Green grapes don't taste green, though, so I know that it does not relate to the fruit's color.

Whereas VE primarily experiences colored hearing, she is one of a handful of individuals for whom synesthesia is bidirectional—smells are colored, and strong colors evoke smells. For example, bright yellow is lemony, and navy blue salty. Note the semantic influence. She also demonstrates the intimate relation between smell and taste. "You know how strychnine smells pink?" she asked matter-of-factly, assuming it should be obvious. "The funny thing is that it *smells* the same pink as the *taste* of my angel food cake. Isn't that remarkable?" Christy Reed also smells colors when handling colored marker pens or cans of paint: "When I see an open can of paint it makes me hungry. I almost want to eat it."

The reverse synesthesia of color → taste also occurs, albeit infrequently (about 2% of self-reported cases). It may have a semantic influence. For one young boy, for example, light pink tastes of a mixture of strawberries and red apples, whereas darker pink evokes a strong taste of strawberries alone.

There is no evidence that novelist Joris-Karl Huysmans (1848–1907) was synesthetic, but his protagonist, des Esseintes, in *Against Nature* (*A Rebours*) composes symphonies with a collection of liqueurs:

Each and every liqueur, in his opinion, corresponded in taste to the sound of a particular instrument. Dry curaçao, for instance, was like the clarinet with its piercing, velvety note; kummel like the oboe with its sonorous, nasal timbre; crème de menthe and anisette like the flute, at once sweet and tart, soft, and shrill. To complete the orchestra there was kirsch, blowing a wild trumpet blast; gin and whisky raising the roof of the mouth with the blare of their cornets and trombones; marc–brandy matching the tubas with its deafening din; while peals of thunder came from the cymbal and the bass drum, which arak and mastic were banging and beating with all their might.

Synesthetic Flavor Responses

"You know why they have music in restaurants?" the synesthete Shereshevsky rhetorically asked Dr. Luria.[21] "Because it changes the taste of everything. If you select the right kind of music everything tastes good. Surely people who work in restaurants know this . . ." Here is just one example of synesthesia *causing* tastes and odors.

Shereshevsky experienced taste in response to multiple stimuli ("Here's this fence. It has such a salty taste and feels so rough . . ."), but he especially responded to sounds and words. Presented with a tone of 50 Hz, he saw "a brown strip against a dark background that had red tongue-like edges. The taste . . . was like that of sweet and sour borscht. . . ." When the pitch was raised to 2,000 Hz, he reported, "It looks something like fireworks tinged with a red–pink hue . . . and it has an ugly taste—rather like that of a briny pickle."[22] Listening to musical works, "I feel the taste of them on my tongue; if I can't I don't understand the music."[23] Describing the shapes of the alphabet to Luria, Shereshevsky said, "I also experience a taste from each sound," indicating that he experienced phoneme–tastes in addition to his fivefold sensory synesthesia. He reported,

I decide what I'm going to eat according to . . . the sound of the word. It's silly to say mayonnaise tastes good. The z (as in the Russian spelling) ruins the taste—it's not an appetizing sound.[24]

For polymodal synesthete Chris Fox, both vision and hearing trigger smells. "Most things I see or hear have a strong taste and smell," he says. For him, letters and numbers possess color, smell, gender, and personality whereas sights and sounds evoke sensations of taste, touch, shape, and color.

Cathleen S is a doctoral student of music. Tastes and smells arise whenever she plays the oboe or piano. This only happens when she herself plays and not when she listens to others play (suggesting that the trigger might be movement rather than sound). Sometimes the tastes and smells are so overwhelming that they interfere with her musical concentration and force her to stop. This distresses her, as does the implication for her career as a performing artist because too often her perceptions are vile and nauseating.

Cathleen's experience is reminiscent of the gustatory synesthesia cited by Schultze[25] in 1912 regarding a case of colored instrumental music. Hearing triggered a sensation first of taste and then of color as if the stimulus "went from the ear through the mouth to the eye." Minor chords caused a hard and bitter taste, whereas major ones were sweet. The difference was sufficiently clear that their subject could accurately identify given musical keys. He spoke of having a "mouthful of music," and reported that both taste and color lingered after the music ceased. Accordingly, he spoke of "digesting" the music.

Recently, a Swiss team examined a 27-year-old female musician referred to as ES[26] who experiences distinctive tastes in response to musical intervals (see table 6.4). Her gustatory experiences are so distinctive that she claims

Table 6.4
Taste responses to musical intervals

Tone interval	Taste experienced
Minor second	Sour
Major second	Bitter
Minor third	Salty
Major third	Sweet
Fourth	Mown grass
Tritone	Disgust
Fifth	Pure water
Minor sixth	Cream
Major sixth	Low-fat cream
Minor seventh	Bitter
Major seventh	Sour
Octave	No taste

(From Beeli et al. [2005].)

to be able to discriminate intervals with extraordinary precision based on the taste generated in her mouth. To test this, the researchers invented a gustatory version of the Stroop task (in which subjects are slower to respond in the face of conflicting stimuli; see chapter 2). They played tones to ES while putting different solutions—sour, bitter, salty, and sweet—on her tongue. Her ability to identify tone intervals, which was perfect in every instance, was significantly faster when the taste applied to her tongue matched her synesthetic taste. ES's ability to use gustatory sensations to identify tone intervals is yet another demonstration that synesthesia can be successfully parlayed into solving complex cognitive tasks.

For Gina P, touching or picking up an object triggers taste. "Velvet isn't so bad if it is velvet on both sides." The tastes are generally not unpleasant, she adds, "except for terry cloth, but that is obvious."

In the case of gustatory phonemes mentioned earlier and expanded below, it is usually the phonetic or, less often, semantic properties of words that trigger taste. However, there appear to be instances in which the *acoustic properties* alone of speech—such as the pitch and timbre of someone's voice—determine the taste. For example, Sue Knifsend remarks that "some people talk like foods." That is, not everyone's voice induces flavors, but those that do are consistent:

. . . in other words, they always talk like the same food. On top of that it usually isn't a normal food. For example, one guy I know talks like barbequed pancakes (are you laughing yet?)—another person talks like spaghetti with M&M's in it (I know—yuck). A woman I've known 10 years always talks like peanut butter.

In this case, speech sounds appear to be cross-activating her gustatory cortex and causing atypical sensory combinations. In a similar vein, Kristen Dolby and her father both taste words—except in them the graphemes rather than phonemes dominate. "Often, the spelling affects the taste. 'Lori' tastes like a pencil eraser, but 'Laurie' tastes lemony. Go figure." In the strictest linguistic sense the semantic meanings of the feminine names are not the same; nonetheless, it is unexpected that graphemes determine a particular taste in this family. Kristen explains:

Some words are a complete "experience" in that they have flavor, texture, tempera-ture, and are sensed in a certain place in my mouth, i.e., back of throat, tip of tongue, etc. . . . Richard tastes like a chocolate bar, warm and melting on my tongue.

Obviously synesthetic "rules"—such as "Graphemes determine color" and "Phonemes determine tastes"—are not absolute but instead good general-izations. In fact, the broad surface variability of synesthetic expression shows it is likely to be explained by ever more fundamental rules such as those operating at the level of gene expression.

Tasted Phonemes

From the largely descriptive character we have given of synesthetic tastes and smells, it is obvious that there has been little modern scientific inquiry into these types of synesthesia. Neurologically speaking, taste and smell are "neglected senses," simply not considered sexy enough or as interesting as vision, for example (their primary cortical areas were shown in figures 6.1 and 6.2; taste is also determined by nuclei in the brainstem). Fortu-nately, this neglect is subsiding. In particular, psychologists versed in lin-guistics such as Julia Simner in Edinburgh and Jamie Ward in London are teasing out the learned linguistic elements in individuals who taste pho-nemes.[27] This is not an easy undertaking.

Prior to their interest, reports of taste as a synesthetic experience were rare—only three cases existed in the historical literature. The Englishman

James Wannerton was the first phoneme → taste synesthete who they studied in depth.[28] He was featured in the BBC *Horizon* documentary, "Derek Tastes of Earwax"[29] and has undergone fMRI scanning. While listening to words, but not tones, James shows a higher-than-normal activation of the main taste areas in the cortex. Drs. Ward and Simner find that James's tastes arise in response to his own speech, to other people's speech, to reading, and to inner speech. Tastes persist for a time until they are "overwritten," which makes his synesthesia a nuisance when he is trying to attend to meetings or read. (A similar, if opposite, interference occurred in Luria's Shereshevsky, who said, "If I read when I eat I have a hard time understanding what I'm reading—the taste of the food drowns out the sense.")[30] Interestingly, taste figures prominently in James's dreams. Without question, his synesthesia is intensified by alcohol and reduced by caffeine and sleepiness. He has difficulty with arithmetic and right–left orientation, and he has eidetic memory. It surprised him to learn that the latter was unusual.

An amusing but predictable observation follows as a result of James's synesthesia going in only one direction: Drs. Simner and Ward have quickly learned many of his taste associations by conventional memory and they can recall the details of his synesthesia in either direction. But James cannot. He can only describe the taste of a given trigger word and not the reverse, a shortcoming he finds funny. This happens because synesthesia is *perceptual* rather than memory based. We examine his kind of perceptions in detail below, but we first summarize what we already know about phoneme tastes.

Briefly, synesthetic tastes are located in the mouth and occur for both spoken and written words. Tastes are highly specific and detailed rather than being basic tastes such as merely sweet or salty. Common words are more likely to elicit tastes than infrequent ones, and actual words (lexemes) are far more likely to evoke tastes than made-up nonwords. Also there is no first-letter effect: words sharing the same first letter do not necessarily produce the same taste. Instead, words containing similar patterns of phoneme sounds (rather than written graphemes) tend to elicit the same taste (e.g., "thriller" and "hillock" both elicit the taste of vanilla). Subsequently, words that share either *sound* ("tangible" = tangerine) or *meaning* ("kisses" = Hershey's kisses) with food words may acquire the corresponding synesthetic taste. All these observations indicate

a need to better understand the interplay between inborn and environmental factors.

Jamie Ward and Julia Simner analyzed 524 word–taste mappings, finding
fifty-nine tastes that occurred with three or more different words (account-
ing for 84% of James's repertoire). Using statistical analysis, they determined the critical phonemes that accounted for each taste, for example,
m = cake and k = biscuit (cracker). A sample is shown in table 6.5.

They confirmed that it actually is the phoneme sound rather than its
written form that determines synesthetic taste. For example the sound of
/g/ tastes like yogurt whether it is expressed as g as in "begin" or x as in
"exactly." Likewise, the /k/ sound tastes like eggs whether it appears as c

Table 6.5
Critical phoneme triggers for JW's synesthetic tastes

Taste	Critical phoneme	Example words
Apple	p	Parents, deploy
Beans, baked	b, i	Maybe, been
Bread	ɾ, aj, ʌ	Enterprise, discuss
Cabbage	g, ɾ	Agree, greed
Carrots	ae, ɾ, s, p aj	Harry, microscope
Coffee	k, ae	Kathy, confess
Cucumber	ju, ə	You, peculiar
Grape	g, ɾ, ej	Grip, great
Jam tart	p, ɑ, t	Partner, department
Jelly	ɛ	Kelly, television
Lettuce	s	Notice, less
Milk, condensed	k, w, aj	Acquire, McQueen
Mint	t, ɾ, u	Truth, control
Onions	ju, aj	Union, society
Peaches	i, f, ʧ	Feature, teach
Potato	l, d, h	Head, London
Sausage	l, ʤ	College, message
Sherbet	f	Lift, fushia
Toast	ou, s, t,	Most, still
Tomato soup	s, p	Super, peace
Tomato	s, ou	So, Sandra
Vegetables	d, n	Earned, owner
Yoghurt	g	Argue, begin

(From Ward and Simner [2003], with permission.)

as in "accept," ck as in "check," x as in "sex," or k as in "fork." Some phoneme sounds can be pronounced more than one way, a trait called allophony. For example long /l/ ("bell," "loop") and short /l/ ("let," "also") variants produce distinct tastes. However, homonyms—words that sound alike but have different meanings—can sometimes override sound in determining synesthetic tastes. Thus, "sea" tastes of seawater, "see" tastes of baked beans, and the letter "c" is tasteless.

Although there is an innate component to taste synesthesia, it is heavily influenced by learned vocabulary and conceptual knowledge. In the case of food words, there is a total correspondence between semantics (meaning) and phonology (sound) of the trigger word and its synesthetic taste (e.g., "rice" tastes like rice, and "onion" like onion). We may thus posit a new rule that words sharing sound or meaning with food words may take on the corresponding synesthetic taste. But how?

In 2005 Jamie and Julia were able to recruit fourteen new individuals who tasted phonemes (only one reported smells, too). Participants were both consistent and reliable in test–retest conditions spanning five months. Unlike grapheme → color synesthetes who detect no variation in intensity from one item to the next, taste synesthetes did perceive intensity differences. Again, word frequency was the explanation: less frequent words had *milder* tastes than high-frequency words, and foreign words or nonwords, when they did elicit a taste, were the mildest of all (e.g., the German word "*einst*" = a little something salty).[31] This suggests there is a developmental history at work as vocabulary and life experience mature together given that someone's precise pattern of taste synesthesia is a result of both heredity and experience. It was fascinating to discover that James Wannerton's diet plays a role in his synesthesia. The more often a food appears in his diet, the more likely it appears as synesthesia. Moreover, his present synesthetic tastes more strongly reflect his childhood diet than his adult food choices.

Linguists know that the first letter of written words holds a special status. It is less visually crowded and thus the quickest to identify, and it may act as part of a reading access code by which graphemes are converted to their sound representations. Accordingly, synesthetic word color tends to be determined by the first letter. The fact this does not happen with phoneme tastes tells us that a completely different mechanism is at work. Unlike grapheme synesthesia, which is tied to the neuropsychology of written word recognition, gustatory synesthesia is not. A different mechanism must be at work during vocabulary acquisition.

Whether or not a taste is present highly depends on word frequency and lexicality; what that taste will specifically then be depends on the sound (phonological) properties of the triggering word. We noted that phonemes clustering in the taste name, such as /idg/ in sausage, also cluster in trigger words for that taste (e.g., "village," "college," and "message" all taste like sausage). The reason this sound-based type of synesthesia occurs both in written and spoken language is that sound codes are neurologically activated during the comprehension of written as well as spoken language.[32]

Furthermore, there is no equivalent of the alien color effect[33] by which, for example, the word "red" elicits a blue color. That is, there are no instances in which a food word has a contradictory food taste (e.g., "milk" tasting like marmalade). However, just as it is possible to induce Stroop interference in grapheme synesthesia, real taste stimuli interfere with synesthetic tastes. For example, chewing mint gum slows down response times compared to no gum while listening to words known to generate tastes (compared to words that do not generate any tastes).[34]

We might ask if it is just coincidence that "dogma" tastes of hot dogs and "Jackson" of cracker jacks. Statistical methods comparing the co-occurrence of phonemes in both trigger words and target words proves that the pairings are far from random. We have no good idea yet how overlapping word sounds link up (e.g., why "Cincinnati" = cinnamon rolls). There are linguistic theories of how the vocabulary of semantic categories such as food are acquired. At some stage vocabulary is linked to a representation of the corresponding taste. How then do additional like-sounding words become coupled to that taste? A link could either be mediated by the taste name or more directly link up from the taste representation to the lexical entry of the phonological (sound) form.

It may be that James Wannerton and similar synesthetes experienced a more sensuous nonlinguistic kind of synesthesia before they were old enough to learn the names of what they ate (which according to table 2.2 occurs around eighteen to thirty-six months of age). It then may have evolved into the form they now have as adults. We have already cited instances of synesthesia changing somewhat over time—especially during childhood and at puberty. This conjecture is supported by the observation that trigger words produce highly detailed descriptions whereas nonwords and foreign words yield generic tastes at best. For example the nonword "bik" = something stiff and brittle, whereas the French "*une*" = something sour and juicy. Whether a word triggers a robust synesthesia depends on whether it has been incorporated into the child's mental lexicon.

Hubbard and Ramachandran proposed[35] that graphemes and colors joined together frequently because their cortical maps lie adjacent to one another on the left fusiform gyrus (recall figure 1.2). In chapter 9 we consider the argument that coactivation of adjacent brain regions depends on the failure to get rid of juvenile connections as the brain matures. However, note that the part of the cortex involved in taste is not situated anywhere near the region involved in graphemes, yet we find the Dolby family above who taste graphemes. Similarly, the areas that house words and their meanings are nowhere near the taste centers, and yet they can trigger synesthetic taste.

Accordingly we do not think the proximity argument is a necessary part of the story; instead, one only needs projections between areas, and such long-range projections are plentiful in the brain. Colored taste is also problematic for the proximity story because taste areas and color regions are likewise not adjacent. The only possible defense for the proximity case is the detail that secondary and even tertiary maps exist for many functions—for example, taste is also represented in several other scattered areas.[36] Accordingly, there remain a few candidate speech and language areas whose participation in tasted phonemes could explain the importance of words and their meaning in this type of synesthesia. We believe the brain to be less "modular" and heterogeneously more interconnected than conventional wisdom admits. We take up that argument in chapter 9.

7 Auras, Orgasms, and Nervous Peaches

An unusual kind of synesthesia exists wherein subjects see colored outlines or auras around people and objects. What makes it unusual is that the precipitating stimulus is not readily obvious as it is in other kinds of synesthesia. In the first edition of his synesthesia textbook in 1989, Richard called the experience "simple" synesthesia because it was the most elementary response he had encountered.

It was essentially the sensing of color alone without any of the other characteristics typical of photisms. For example, Bruce Brydon explained how he perceives "additional colors" outlining objects, sometimes washing over them in "soft splotches."[1] While sensing the colors, he simultaneously feels emotion-based sensations such as numbness, flushing, exhilaration, fear, or happiness.

The colors he sees are primarily green, red, brown, and amber, either singly or mixed. They generally outline objects such as the Golden Gate Bridge. Although the bridge is actually painted orange, Bruce sees it covered in a green hazy outline. However, the bridge's color can change; it is not always green, an apparent variability that only adds to his synesthesia's strangeness. At other times aura colors can be so intense as to obscure real objects. For example, turning to face someone and "not seeing her but a small white center surrounded by black and a large green masking any feature of this person." It is usual for affective colors to superimpose themselves on the visual scene.

Today we call the phenomenon "emotionally mediated" synesthesia because it turns out that the emotional meaningfulness and degree of personal familiarity something has to the perceiver determines whether or not it triggers a synesthetic aura. In retrospect, Bruce had perfectly described the heightened emotion and portentousness accompanying an aura even as he remained baffled as to its cause:

One particular person comes to mind. There was a very strong feeling and she was surrounded by a dark blue–green aura. It wasn't because she was sexually attractive. I don't know what the emotional feeling was due to because I had only met her twice. But there it was. I think there was some sort of bond or something. I'm not sure which comes first, sometimes I think I see the color and react emotionally; others it may be reversed—I get an emotion and then see this color.

The variability in how Bruce perceives the Golden Gate Bridge (e.g., as green one week then amber the next) violates the general rule that synesthetic percepts are stable over time—but only if we mistakenly think that the physical bridge is the stimulus triggering the aura he sees. It isn't. Rather, personal acquaintances and objects like the bridge induce a color experience by virtue of the emotional valence they arouse *in the receiver*. In other words, Bruce's own emotional judgments trigger the auras.

This counterintuitive setup refers back to early stages of how we first become emotionally intelligent. Part of it requires identifying emotion in ourselves and "reading" it in others. Every self-conscious individual has a "theory of mind" that other individuals have minds of their own. This prompts individuals to discern the other's intent. We figure out what others mean to do largely by reading their emotions, which includes body language, tone of voice, and other implicit behavior. Reading emotion is always a two-way street between a sender and receiver, and because the exercise is not explicit it is always possible to misread another's intent.

We learn to read others before we learn to speak, although the skill is an ongoing learning process that some become quite good at. Between twenty-four and thirty months of age we also learn the primary colors—red, blue, and green. Thus, Angie[2] is scientifically plausible when she says this:

My 4-year-old son also sees people in colors, and has done since he was two. Ever since he knew the names for colors, he could connect them verbally with people. Generally, the people closest to him have the boldest, most definite colors, whereas the people he does not know well *do not yet have colors*. These assigned colors remain unchanged. He also likes to give people things in "their" colors [*italics added*].

Note how color intensity relates to emotional intensity, with strangers usually lacking any color at all. This phenomenon is echoed in the literature. For example, the clinical description of a seven-year-old indicates that aura color is determined by the degree of acquaintance rather than physical characteristics or the acquaintance's mood. Thus, plaster busts elicit no color, yet all real strangers are described as being "bright orange with a black outline. . . . As I know them better they get mild blue or pinkish–orchid. . . . When I know people well they stop changing colours; they are the colour."[3]

We categorize others favorably or not and also judge their personalities while we make up our minds about them. We calculate such attitudes quickly and automatically. And once we "size them up," our attitudes tend to remain fixed. Thus, it is no surprise that color auras are in fact stable over time once the emotional attitudes of the perceiver settle down. As Cameron La Follette notes,[4] children's colors are generally "pale and insubstantial until their personalities are more settled, generally by age 12 or 13." He finds that children's colors deepen and solidify as their personalities develop. After that, people's colors

stay the same always; it is not related to their moods or other things. Color is related to personality because, for example, pink people always have some childhood emotional problem which they are still living; yellow people are extroverted and fun to be with; and so on. "Bubbly" people literally have bubbles in their synesthetic color. For me, texture is almost equally important as the color.

He notes that animals can have auras too but, most importantly, "I have to know an animal's personality to know its color." For him, animals are monochromatic whereas human colors tend to be striped, spotted, or textured, which makes sense to him because animal emotion is presumably less nuanced than human emotion.

Just as children's insubstantial colors deepen as their personalities mature, the opposite happens when individuals fall apart at life's end. Individuals with emotionally mediated synesthesia have sometimes claimed that color loss was correlated with illness or death. Alzheimer's patients—who act emotionally blank or revert to emotionally immature states—have "washed-out and runny colors" according to Cameron La Follette, whereas Elizabeth N. ran into an acquaintance who no longer triggered any color experience—and who passed away shortly thereafter.[5] Needless to say, this report does not indicate clairvoyance, but instead that one's interpretation of the health of another, whether conscious or not, can influence the colors.

Usually synesthetes do not explicitly make the connection between how *they* read a person and the auras they perceive. Deni Simon, for example, relates how new acquaintances who have a "color spike" around them invariably turn out to be disagreeable. When she meets someone surrounded by "zigzag static . . . small lightning bolts, metallic, silvery, bronzy," she just doesn't like them.

For rare individuals, however, the emotional connotation is overt. "Instead of seeing a person as mad, sad, or sick," says Marnie Loomis, "I see a person who is red, green, or black. I often see shapes around them as well, and use this information when working with patients"[6] (she is a naturopath). However, note that even while she identifies color with emotion, she still mistakenly attributes it as *coming from* the other rather than originating in her own reading of that person. We would argue that there is nothing special about synesthetic auras other than their being an overt marker for a process that is normally unconscious and automatic, namely, reading other persons.

Jamie Ward conducted several experiments to prove that emotional connotation is indeed the fundamental trigger for auras.[7] For example, compared to other words, English Christian names are a good inducer of color in individuals with emotionally mediated synesthesia. Ward showed that names of people his subject personally knew were far more likely to induce color than names referring to persons with whom she was not acquainted (see figure 7.1). This is because such names are more likely than nonreferring ones to elicit an emotional response. Consistent with this, other word categories including color names themselves did not elicit synesthetic colors whereas emotionally loaded words (e.g., "love," "anger") did tend to do so (see figure 7.2). Variables such as word frequency and imageability were controlled for, leaving emotional connotation as the likely explanation. Ward's subject was far more consistent in a test–retest situation over time compared to controls and, as expected, showed Stroop interference during naming. Both results indicate that auras are generated automatically.

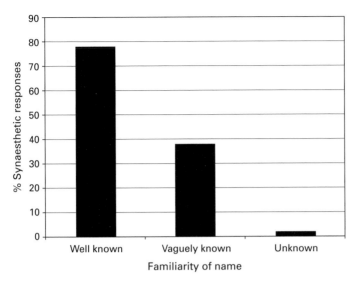

Figure 7.1
Influence of name familiarity on the probability of eliciting a synesthetic response.
(Adapted from Ward [2004].)

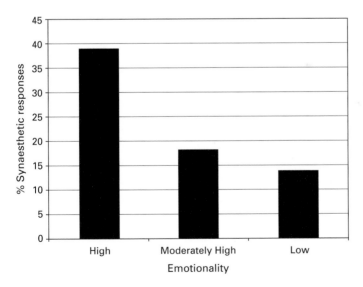

Figure 7.2
The more emotional common nouns are, the greater the probability of their eliciting
synesthesia. (Adapted from Ward [2004].)

The fact that personal familiarity affects synesthetic color is another example of how life experience shapes synesthetic correspondences, which are typically considered innate. Valence (positive or negative) may be the means by which new acquaintances acquire synesthetic color in individuals predisposed to auras. Tellingly, Ward's subject experiences color only when presented with people's *faces*, as did a patient of Ramachandran and Hubbard.[8] Faces are well-known elicitors of emotion, as demonstrated by the large galvanic (electrical) skin resistance they evoke.

Emotionally mediated synesthesia appears to be rare. In addition to Cytowic's case and his own, Jamie Ward found only six others reported in the literature. There are too few cases to know whether color and emotion pair up in any regular way reminiscent of the Berlin and Kay order for color naming. However, anthropologists have found consistent color–emotion associations between Western cultures and ones with minimal Western contact.[9] For example, darker and less saturated colors are associated with negative emotions whereas lighter and more saturated colors are associated with positive ones. Importantly, and in keeping with earlier comments about equivalences across sensory dimensions, it is aspects within color space (e.g., brightness, saturation) rather than hue per se that underlie cross-sensory associations. What the anthropological work shows is that acquired knowledge about people can tap into intrinsic and universal mechanisms.

Regarding possible components of a neural network for colored auras, we can only speculate. One possible node is the retrosplenial cortex, an area that responds both to familiar and to unfamiliar people, and to emotional words relative to neutral ones.[10] It is active during emotion and memory tasks.[11] Neuroimaging data supports a role for this brain area: in one study of a synesthete who experiences colors in familiar people, activation was seen both in color areas and in the retrosplenial cortex.[12]

The claim that certain individuals perceive colored auras holds an important place in folk psychology. In some instances, clear parallels are evident with scientific reports during the last century. Although the bulk of people claiming to have such power are charlatans, it is possible that some may be undiagnosed synesthetes. Rather than assume that people radiate energy only adept seers can detect, a scientific account need only assume that people regularly elicit an emotional response in the perceiver, and that a synesthetic cross-activation between brain structures involved in emotion and color perception enables an emotion-inducing stimulus to explicitly acquire the novel aspect of color. The result is a color aura.

In its purest form, emotionally mediated synesthesia triggers only color—but as the above examples indicate, other cognitive properties such as texture, shape, and movement may be part of the elicited experience.

Synesthetic Orgasms

If ordinary emotional valence can induce color, shape, and texture, then what about that most intense and paroxysmal emotional experience, the orgasm? Though we have not systematically queried individuals about what happens during sex, a fair number have volunteered comments that orgasm indeed causes color, shape, texture, movement, and taste. As Susan Meehan explains:

My favorite orgasms are brown, 2-dimensional squares, which I know doesn't sound very exciting, but they are extremely pleasurable. Naturally I have other colors and shapes, but these are particularly wonderful. Of course, my husband doesn't understand at all, but he's pleased when I get them.

"My husband likes to hear about the colors," is a common refrain. Orgasmic photisms are variously described as "brilliant flashes of colored lights," or "two dimensional, brightly colored shapes moving against a black background," or "neon pastels, intertwined like a rope or thick strands of licorice," or like "a three-dimensional oil slick, myriad colors blended together exactly as it looks on the road after a rain."

One's first orgasm is generally an astonishing event. Sometimes, it produces a one-time synesthesia, which is what happened to Dmitri Nabokov:

As a very young teenager, my sexual awakening and my first intense sexual experience was accompanied by huge, strong, geometrical shapes, spheres, cubes, and pylons that filled my mind and that never returned.

As with aura experience, the key to synesthetic orgasms is emotion. And whereas the emotional judgments that trigger an aura are typically covert, explicit feelings can trigger one too. As Deni Simon explains:

People say that anger is red, but I see purple. If I'm really upset at my kids and I'm yelling at them, there will be a purple background behind them. Where their head meets the background, it's like an aura, a yellow luminescence going into purple. It dissipates slowly.

It should therefore come as no surprise that orgasms trigger synesthesia given that an orgasm is a massive emotional discharge. Orgasm involves a widespread discharge of activity throughout the brain, resulting in both implicit and explicitly felt components.[13]

We need to point out an important distinction that neuroscience makes between feelings and emotion because the two are not identical even though everyday language conflates them. Emotion is unlearned behavior, an unfelt automatic script playing itself out, whereas feeling is the mental readout of emotional scripts. The unlearned reactions of emotion always change the body's physical state—thus, everyday language refers to "gut" feelings, but far more gets activated: the face, the skeletal muscles and overall posture, the breath, heart rate, and viscera, all the way down to the chemical soup of the internal milieu. Because emotions evolved as supervisory routines for homeostasis (that is, keeping the internal milieu stable), emotions relate to the management of life. Emotional scripts play out unconsciously until feeling enters in: only then does the mental readout let us take stock and, if we are emotionally intelligent, make a connection between what we feel and what triggered it.

As we stated, the manifestations of orgasm are both implicit and explicit. Actually, the human sexual response in both men and women has four phases: arousal, plateau, climax (orgasm proper), and resolution. Inquiry into synesthetic orgasm is insufficiently advanced to permit knowing whether synesthesia occurs only during the climax phase, although that is the phase of maximal discharge from limbic and hypothalamic nuclei, or throughout its stages. (For obvious reasons we have excluded synesthetes in whom touch is the primary trigger.)

However, given that orgasm is followed for some time by heightened emotional outflow compared to baseline, we would predict lingering synesthesia during resolution and afterwards. The experience of Karin S fits this prediction: after climax, her "colors remain a while afterwards . . . and it takes some time, maybe four or five minutes, before I can see clearly again."

In some individuals kissing is another reliable synesthetic trigger that has implicit and explicit emotional aspects. Deni Simon's personal experience inspired the title of the BBC documentary "Orange Sherbet Kisses":

Pain and pleasure sensations evoke visual/spatial perceptions which are also in color. In fact, I was recommended for psychological counseling in high school because I told the assistant principal that when I kissed my boyfriend I saw orange sherbet foam.

Of course, it is impossible at this time to distinguish whether the synesthesia is triggered by the tactile input from the lips, the limbic (emotional) activation, or the cognitive states(s) associated with the many kinds of kisses. At this stage we need more first-person accounts from such synesthetes. Incidentally, all our reports of synesthetic orgasms thus far are from women. It is possible that men do not experience this; alternatively, the absence of male reports may result from the observation that women more easily self-disclose.

Lastly, we have only once encountered the reverse situation, namely, erogenous feelings as a synesthetic response. Michael Watson, the man who tasted shapes, literally had his libido supercharged by taste stimuli. Novel flavors, such as those encountered in restaurants, particularly triggered libidinous feelings as he recounted during dinner at Maison Blanche in Washington one evening:

Eating for me is an impulse and the first bite I take of the new course is an urge to look into a new direction. I feel drawn by this. These new experiences are frequently very erotic, in the term of sensual, although sometimes they are erogenous as well and there are times when I am eating when I just want to push the table over and screw whoever is nearby.[14]

Nervous Peaches and Unusual Experiences

Just as some synesthetes personify graphemes, so others project emotional traits onto inanimate objects. For example, Susan Meehan says, "I know this sounds completely absurd, but the other week my husband and I were in the produce section of the market, when I grabbed his arm and said, 'I don't know why, but those peaches are extremely nervous.'" Another synesthete hesitates to tear a banana from the bunch because it will then be "lonely." What is going on?

It is hard to say what causes this misattribution of affect. Although synesthetes report the phenomenon often enough, we do not regard it as a manifestation of synesthesia *per se* because the defining elements of trigger and response are missing. Rather, we regard it as an epiphenomenon, or consequence, of having a synesthetic brain. The misattribution is a cognitive error we include within the boarder phenomenal category of unusual experiences.

We cannot always take such experiences at face value. However, we can attempt a neurological explanation for them. If some synesthetes have either a higher baseline of limbic tone or a hyperconnectivity between limbic structures and other brain areas,[15] then their experiences might make sense given that similar ones are seen in recognized neurological conditions.

For example, psychic experiences have long been associated with seizures, especially the temporal–limbic variety.[16] Experiential manifestations may include a sense of portentousness or other affective state, feelings of unreality, out-of-body sensations (autoscopy), forced thinking, memory flashbacks, and illusions of familiarity or unfamiliarity called déjà vu or déjà vécu.[17] The latter terms mean "already seen" or "already experienced" and convey the sense that what the patient is experiencing at the moment has been witnessed or lived through previously. This often leads individuals to conclude they are clairvoyant or prescient ("I must have known it was going to happen because it was all familiar to me").[18]

There is of course a difference between the feeling of knowing and actually knowing, which individuals often do not bother to distinguish. Moreover, humans are notoriously poor statisticians. We focus on those events we feel are significant but ignore prior predictions that turned out differently. In other words, we tend to overestimate what is significant by ignoring the ordinary.

Consider the feeling of a presence. Penny P is a synesthete who experiences blobs of color that either "visit" or "help" her. For example, she reports that a translucent but not transparent red blob covered the back of her writing hand while she completed a difficult examination; another time a warm blue light hovered about her left arm and shoulder as if the sun were shining on it as she wrote a letter on an emotional subject. Her "visiting colors" include a purple three inch by four inch oval that appears daily and a smaller blue light than often appears when she puts her baby to bed. "I've thought of these as angels, but who knows what they are," she says. Note the presence of heightened emotion in all her scenarios.

The feeling of a presence or the "visitor experience" is known to clinical neurology and correlates with changes involving the midline and bottom (amygdaloid–hippocampal) parts of the temporal lobes.[19] These brain areas are associated with, among other things, the experience of meaningfulness, the sense of self and its relationship to space–time (including religious and cosmic associations), dreamy states, feelings of movement, and smell.[20] Accordingly, affected individuals report sensing a presence, feelings of floating or hovering, and flashes of objects "just missed" out of the corner of their eye. The latter occurs because the most peripheral part of the visual field projects to the salient parts of the temporal lobe.

Regarding the nervous peaches and other examples of misplaced affect that started our discussion, "psychic seizures" refer to epileptic discharges that induce physical sensations, emotional affects, and trains of thought (forced thinking) without causing a shaking convulsion.[21] We have known for a long time that repeated seizure activity is likely to cause "kindling," meaning a hyperconnectivity between different brain regions. Kindling has a genetic factor whereby repeated stimulation of the brain at levels initially incapable of eliciting convulsions induces a permanent susceptibility to seizures. When seizures and kindling strengthen sensory–amygdala connections, for example, patients subsequently experience heightened emotions in response to specific sensory inputs,[22] which become increasingly meaningful[23] (e.g., blue takes on special significance). Accordingly, if temporal–limbic structures in synesthetes' brains are hyperconnected, synesthetes would then be capable of suddenly feeling an emotion that "they" didn't cause. Arising out of nowhere but needing explanation, the emotion becomes misattributed to an external object.[24] Psychologically, humans have a need for such closure even if the conclusion is absurd, as it is in the case of nervous peaches. Compared to the baseline population, synesthetes may simply have what is called a *widened* affect.

8 Metaphor, Art, and Creativity

In this chapter we ask three related questions. How can synesthesia help us better understand the neurological basis of metaphor, what kinds of art does synesthesia inspire, and what might synesthesia tell us about creativity? As we shall see, each of these issues depends on making connections across concepts in the brain.

Making Sense of Metaphor

As Ramachandran and Hubbard pointed out in 2001, metaphors often take the form of cross-sensory associations such as "loud tie," "cool jazz," "sharp cheese," and "sour note."[1] But why do we have such synesthetic turns of phrase and why is their meaning so readily obvious? For example, when we describe people as "sweet," we do not literally mean that if we were to lick them like an ice-cream cone they would taste sweet. Rather, we mean they are pleasant and agreeable the way sweets are. Schizophrenics actually do interpret metaphors literally, while the rest of us understand them with no problem. Why is that?

We already indicated that synesthetes and nonsynesthetes both make matches in ways that are similar. The work of Larry Marks and others unmasks systematic and lawful correspondences among sensory dimensions such as size, pitch, lightness, loudness, visual position, and shape[2] (see figure 8.1). For example, both groups say that louder tones are brighter than soft tones, higher ones are smaller and brighter than low ones, and low tones are both larger and darker than high ones. Even smell maps to bright–dark and high–low dimensions, the relation between color, lightness, and odor intensity being well-known to both psychologists and chefs: for example, people agree that a darkly tinted liquid smells stronger than

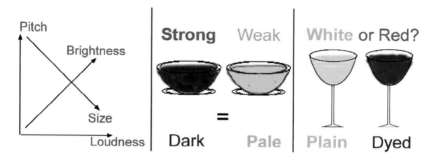

Figure 8.1

(*Left*) Just as there are lawful and regular sight–sound correspondences among pitch, size, loudness, and brightness such that the metaphor "loud shirt" makes eminent sense, color intensity maps to the strength of taste and smell (*center*) such that darkly tinted liquids are judged to be stronger than pale ones. Additionally, color hue can prime specific aromas (*right*). In this example, white wine surreptitiously dyed red is said to smell like red wine.

when it is pale. A darker color also makes them say that the taste is stronger[3] (see figure 8.1, *center*). Similarly, when white wine is surreptitiously colored red it suddenly smells like red wine, an example of color priming (see figure 8.1, *right*).

We pointed out how less obvious dimensions of perception map to one another seamlessly. For example, sight, sound, and movement map to one another so closely that even bad ventriloquists have no trouble making us believe the moving dummy is talking (see figure 8.2). Dance was another example we gave of cross-sensory mapping in which body rhythms effortlessly correspond to sound rhythms both kinetically and visually.

We now wish to show that the capacity for cross-sensory mapping underlies the ability to make metaphors—by definition, the ability to see the similar in the dissimilar.

Let's begin with the assumption that the gene(s) for synesthesia cause normally unconnected brain areas to interact with one another, thus linking apparently unrelated qualities such as sound and color or November and a spatial location. Now suppose that this gene were expressed in more than one area in the brain. Ramachandran and Hubbard suggested that this could, in theory, allow a generalized hyperconnectivity enhancing the ability to link seemingly unrelated things—which is the hallmark of both metaphor and creativity.[4] Of course, there is more to creativity than

Figure 8.2
Sight, sound, and movement are inherently tightly coupled, which is why even bad ventriloquists convince us the moving dummy is talking. The ability to cross-link different *sensory* dimensions underlies our ability to make *conceptual* metaphors.

a capacity for metaphor. Creative people easily combine parts into a whole, or identify parts in a whole then shift from one whole to another. They are open to stimuli, have a wide span of interests, and are unconcerned with social norms. They also have a high degree of self-acceptance, independence, flexibility, and enthusiasm and can suspend the constraints of the intellect on their imagination. Nonetheless, could increased connectivity between different brain regions be a part of the story?

Consider the following experiment first conducted in 1929 among different cultures.[5] When speakers of various tongues are shown the two shapes in figure 8.3—one looking like a jagged shard of broken glass, and the other like an amoeba blob—and told that in some alien language one of the shapes is called "bouba" and the other "kiki," 98% pick the spiked shape as "kiki" because its visual jags mimic the "kiki" sound and the sharp tongue inflection against the palate. By contrast, the blob's rounded *visual* contours are more like the *sound* and *motor* inflections of "bouba."

This kind of correspondence across cultures illustrates the rule that pre-existing relationships (analogies) are often co-opted in biology. In this way,

Figure 8.3
When speakers of various tongues are told that in some alien language one of these shapes is called "bouba" and the other "kiki," 98% of individuals pick the spiked shape as "kiki" because its visual acuteness mimics the motor and auditory inflections of "kiki" when spoken.

synesthetic associations our ancestors established long ago grew into the more abstract expressions we know today—and this is why metaphors make sense. Orderly relationships among the senses imply a cognitive continuum in which perceptual similarities give way to synesthetic equivalences, which in turn become metaphoric identities, which then merge into the abstractions of language. In other words, the progression looks like this:

perception → synesthesia → metaphor → language

Metaphor is therefore the reverse of what people usually assume. It depends not on some artful ability for abstract language but on our *physical interaction* with a concrete, sensuous world, as Berkeley linguist George Lakoff first demonstrated. Closely following his examples, let us see how our many *orientational metaphors* grow out of the physical experience with UP and DOWN polarities.[6]

Conscious is up; unconscious is down

Wake *up*. I'm *up* already. I'm an early *riser*. I *dropped off* and *fell* asleep. The patient went *under* anesthesia, *sank* into a coma, then *dropped* dead.

Controlling is up; being controlled is down

He's *on top* of the situation, in *high* command, and at the *height* of power in having so many people *under* him. His influence started to *decline*, until he *fell* from power and *landed* as *low man* on the totem pole, back *at the bottom* of the heap.

Good is up; bad is down

High quality work made this a *peak* year and put us *over the top*. Things were looking *up* when the market *bottomed out* and hit an all-time *low*. It's been *downhill* ever since.

Rational is up; emotional is down

My heart *sank* and I was in the *depth* of despair, unable to *rise above* my emotions. I *pulled myself up* from this sorry state and had a *high-level* discussion with my therapist, a *high-minded*, *lofty* individual.

The physical basis for these common metaphors is that most mammals sleep lying down and stand up when awake. Well-being, control, and things characterized as good are therefore all UP. Because we control our physical environment, animals, and sometimes even other people, and because our ability to reason is assumed to bestow such control, CONTROL IS UP implies HUMAN IS UP and therefore RATIONAL IS UP.

Spatial orientations such as up–down, front–back, and center–periphery are the most common ones in our system of concepts, but others exist given the variety of ways we interact with the world. We make inside–out distinctions between reason and emotion, for example, and generally characterize rationality as up, light, and active, while the emotions are down, deep, and murky—passive, irrational passions over which we presumably have little control.

Anthropologists tell us that the major orientations of up–down, in–out, center–periphery, and active–passive exist in all cultures. However, which concepts are most valued varies from place to place. Some cultures prize balance, whereas English-speaking cultures seem taken with the extremes of UP or DOWN orientations.

Whereas interactions with space yield *orientational metaphors*, other experiences give rise to what are called *ontological metaphors*, ways of treating events, actions, emotions, and ideas as reified, self-contained objects. (Ontology is the philosophical investigation of the nature of being.) Cultural influence elaborates ontological metaphors. For example, we can elaborate THE MIND IS AN ENTITY into either THE MIND IS A MACHINE or THE MIND IS A BRITTLE OBJECT to arrive at the following different meanings:

The mind is a machine

We are *cranking out* a lot of paperwork. You could see his *wheels turning*. Their proposal just *ran out of steam*.

Compare this with the results of a different elaboration:

The mind is a brittle object

He *cracked* under pressure. It was a *shattering* experience. You *bruised* his ego. His mind just *snapped*.

Metaphors therefore emphasize some facets of an entity while hiding others. The MACHINE metaphor casts the mind as having a power source, an anticipated level of efficiency and production capacity, and an on–off state. However, it hides the vagaries of thought, the mind's ability to deal with incomplete information, and its subjective leaps of intuition.

Switching metaphors alters how we comprehend a thing and thus alters reality. Words in themselves cannot change reality, but changing from one concept to another does alter what we perceive and how we act on those perceptions. Ontological metaphors are so pervasive that they seem natural and self-evident. It never dawns on us they are metaphors. Consider, for example, the physical experience implicit in the following:

Understanding is seeing; ideas are light

I *see* what you are saying. It was a *brilliant* remark and a *clear* discussion. Your point of *view* gave me the *whole picture*. Their proposal is *murky*, the ideas *opaque*, and their premise *transparent*.

Emotion is physical contact

The verdict *bowled him over*. I was *struck* by his generosity. His donation *made an impression* on me. That model is a *knockout*. I was *touched* by their kindness.

It is evident that different metaphors produce different flavors of a given concept. The intuitive appeal of any concept rests on how well its metaphors agree with our actual experience. Thus, we can elaborate IDEAS ARE

LIGHT more easily than IDEAS ARE SOUND, saying, for example, "He had a *bright* idea" but not "he had a *high-pitched* idea." (The SOUND metaphor is not entirely vacuous, because we might say "He was *buzzing* with ideas," or "He *quieted* his mind.")

One factor contributing to the discord in human minds is the conflict among metaphors that arises from real differences in their physical basis. For example, "That's up in the air," and "The matter is settled" are each physically consistent with "I grasp your meaning." If you can grasp something you can examine and understand it, and things are easier to grasp if they are down rather than flying about. Thus, UNKNOWN IS UP and KNOWN IS DOWN are coherent with UNDERSTANDING IS GRASPING. However, UNKNOWN IS UP is inconsistent with the orientational metaphors GOOD IS UP or FINISHED IS UP. Logic requires that FINISHED be coupled with UNKNOWN, and UNFINISHED with KNOWN. But our experience disagrees! We do not consider the unknown to be good. Thus, the physical experience leading to UNKNOWN IS UP is entirely different from that on which the two incongruent metaphors are based. This may suggest that the ability to be at odds with ourselves or to simultaneously hold conflicting beliefs can be rooted in physical experience rather than faulty logic.

Note that from an early age, creative people have a high tolerance for just such inconsistency, contradiction, and paradox whereas literal-minded people cannot get past such obstacles to fathom a connection. And as for the poor aptitude of schizophrenics, it is precisely because they are literal minded that they cannot grasp even the simplest metaphors. Thus, creativity and the capacity for metaphoric connections go hand in hand. It is a small step, and an earlier one at that, to go from orientational metaphors to ones such as "loud tie" or "sweet person."

Some sensory similarities such as brightness–loudness, which is the physical basis of "loud tie," manifest early in life and appear to be innate. Such similarities produce "synesthetic metaphors." Other similarities such as color–temperature or pitch–size manifest much later, around adolescence, and appear to be based in experience (see figures 8.4 and 8.5). Children as young as four years old recognize perceptual similarities between hearing and vision in that they can match pitch and loudness to brightness. For example, they consistently rate "low pitched" as dim and "high-pitched" as bright.[7] Not until about age 11, however, do children recognize similarities between pitch and size, or reliably recognize "warm" and "cool" colors.[8] We again note that this is the approximate age when synesthesia falls off

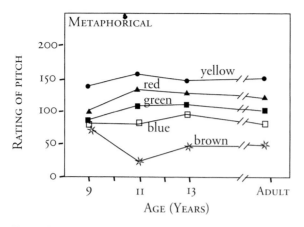

Figure 8.4

There is a striking ordering of five color names with respect to auditory pitch in both children and adults (*N* = 500). (From Marks, Hammeal, and Bornstein [1987], with permission.)

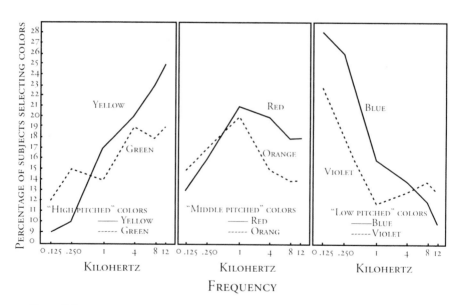

Figure 8.5

Children in grades 3–6 (ages eight to twelve years) consistently match sound pitch with the same colors. (*Left*) "high-pitched" colors; (*middle*) "middle-pitched" colors; (*right*) "low-pitched" colors. (From Simpson, Quinn, and Ausubel [1956].)

sharply. Later, with increasing age, language gains access to this graduated sensory knowledge, thus permitting the interpretation of synesthetic metaphors according to cross-sensory physiology. Improvements with age in making metaphoric translations of synesthetic expressions parallel both the child's increasing differentiation of literal meaning, and an increasing capacity to integrate component meanings into compound expressions.

Whereas innate perceptual correspondences produce synesthetic metaphors, learned associations produce *synesthetic metonymies* (a metonymy is a relationship based on association rather than equivalence, and most are learned). Both help to elaborate the more developed *ontological metaphors* such as HE CAME UP WITH A BRIGHT IDEA. Perceptual synesthesia and metaphoric synesthesia therefore may both emanate from an experienced similarity of perception in different senses.

A question we have not yet answered is how polarities come about in the first place. Synesthetic equivalences provide a starting point for transforming the sense-bombarded newborn into the mature, metaphoric adult as perceptual equivalences make themselves available to language. Once children understand that they can map one sensory pole to a different sense modality (e.g., high–low to bright–dark), they can then extend the process to nonsensory categories, rendering the new experience bipolar (e.g., strong–weak, or positive–negative). The same sort of neurophysiology that underlies sensory dichotomies can thus also mediate more abstract semantic features to which the properties of order, gradation, and polarity also apply. Polarity has special importance because a nonsensory feature can now be represented by two opposing processes with reversed polarities. This leads to the orientational metaphors discussed above such as those based on polarities of UP–DOWN, FAR–NEAR, and IN–OUT.[9]

As Larry Marks[10] pointed out, "many of the very same rules that govern the perception of the synesthetic minority also govern perceptual and verbal behavior of the non-synesthetic majority." This suggests that synesthesia rests on a shared core of cross-sensory similarities—ones that appear to be innate in early life and derive from fundamental similarities in phenomenal experience itself. These in turn gradually become available to the more abstract system of knowledge embodied in language.

Marks[11] concludes that perceptual experiences of meaning are multidimensional and that verbal (semantic) knowledge taps earlier perceptual knowledge. This conclusion is echoed by Sean Day[12] who notes that colored sounds are the most common expression of perceptual synesthesia, whereas metaphoric elaborations of tactile sound are most common in (English) literary synesthesia. It appears likely that human thought itself is largely metaphoric. Hearing is the sense most frequently expanded by both perceptual synesthesia and synesthetic metaphors. Sean Day also concludes that synesthetically seeing sounds, which antedates language, has probably influenced language development.

We noted earlier that synesthesia is more common in children than adults. Based on nineteenth- and early-twentieth-century reports, children are roughly three times more likely than adults to be synesthetic. Still unanswered is why most children lose their synesthesia as they grow up. Perhaps synesthesia is too general, imprecise, and inflexible a form of cognition compared to language and the later-developing meanings that are possible in the verbal realm. Psychology calls this the transition from "iconic" to "symbolic" modes of thought.[13]

To sum up, childhood thought becomes condensed and accelerated as it is internalized from more physically based cross-sensory associations to the felt meaning of inner speech.[14] Using his most famous metaphor, William James[15] spoke of "feelings of relation" in the "stream of consciousness." As color becomes a synonym for, say, a piano key, a grapheme, or a taste, cross-sensory translations inwardly couple the features of each as a feeling of sameness. Synesthesia is about identity rather than associations.

Mid-childhood synesthesia emerges as the differentiation and internalization of earlier cross-modality in the infant—for example, facial mirroring or visual–tactile play with objects. Neonates' visual–gestural mirroring[16] is associated with activation of left frontal and temporal areas that are later associated with language.[17]

Children internalize speech by age six to seven, and felt meaning is synesthetically structured, being the basis of metaphoric understanding that develops later in childhood. The further appearance of synesthesia in adults who score high in traits of imaginative absorption and openness to experience,[18] as well as during meditative and drug-induced states, also supports the notion that inner speech is synesthetically structured.

Synesthetic Art

For a long time it has been thought that synesthesia is more common in artistic individuals. Whether this is statistically true or not is unknown

Table 8.1
Artistry in synesthetes and nonsynesthetes ($N = 51$)

Characteristic	Synesthetes	Nonsynesthetes
Art, design, entertainment occupation	25%	0%
Formal art or music training	85%	15%
Hobbies in art or music	79%	29%
Mean number of languages spoken	3.6	1.6

because no one has systematically examined a random sample for both traits. There are, of course, notably famous synesthetes such as Liszt, Hockney, and Nabokov. However, here our experience is highly skewed because there are far more famous artists who happen to be synesthetic than there are synesthetes who happen to be famous artists. It is again a matter of sampling bias.

Still, the data are suggestive. For example, 41% of subscribers to The Synesthesia List (a nonrandom sample) work in artistic professions. Psychologist Carol Crane finds that synesthetes are far more likely than nonsynesthetes to have artistic or musical training and know a foreign language (see table 8.1).[19] Self-reported nonrandom surveys also come up with a high percentage (23%) of artistically inclined persons among synesthetes[20] (methodological shortcomings in that study, however, may merely indicate that synesthesia is more common among fine art students than the general population).

What kind of art do synesthetes create? Usually they take one of several approaches. Some, like Jane Bowerman, who has [sound → color, shape, movement] + [emotion → color] synesthesia, are self-taught and paint only what they see (see figure 8.6):

I can do a painting and not have a clue about how I did it because I don't know how to draw. I couldn't possibly paint anything my little 3-year-old grandchild can draw. I don't know how to do it. But I do know how to draw what I see.[21]

Another approach is for synesthetes to attempt an exact portrayal of what they perceive, perhaps even providing a guide. Composer Olivier Messiaen, for one, does this when telling us "certain combinations of tone and certain sonorities are bound to certain color combinations, and I employ them to this end."[22] In describing the colored chords of the

Figure 8.6
Photisms painted by Jane Bowerman are strikingly similar to the form constants and replicate individual elements. (*Left*) Repeated circular forms move out from the center. (*Right*) A symmetrical central radiation moves outward; black dots indicate scintillation.

"strophes" that comprise *Chronochromie* (literally, "color of time"), he explains,[23]

one note-value will be linked to a red sonority flecked with blue—another will be linked to a milky white sonorous complex embellished with orange and hemmed with gold—another will use green, orange, and violet in parallel bands—another will be pale gray with green and violet reflections . . .

Not only does he strive to express the precise colors of sound, but he also tells us exactly what they are. Furthermore, referring to the score of *Colours de la Citie Celeste* by which he recreates the luminous effect of stained glass,

I've noted the names of these colors on the score in order to impress this vision on the conductor who will, in his turn, transmit this vision to the players he conducts: the bass should, dare I say it, "play red," the woodwinds should "play blue" etc.[24]

We note that Messiaen has had a lifelong fascination with color and colored light ever since his childhood exposure to medieval stained glass. Depending on a given artist's skills, attempts at direct translation may come across either as naive and ham-fisted or as intrinsically interesting like Messiaen's music, which is stylistically so unique as to be instantly recognizable. Given the idiosyncratic nature of synesthesia, however, whatever special meaning it holds for the artist is certain to escape us no matter how much the synesthete tries to help us "get it" by way of explicit guides.

A different approach is the private reference, for which Vladimir Nabokov is especially known. Aside from inserting overt synesthetic events into his stories, he includes synesthetic equivalences for his own amusement and does not care whether we get the joke or not. For example, consider the title of his sweeping saga *Ada*. Does the title mean anything, or is it merely a proper name? The book itself is circumspect. In Nabokov's colored alphabet, however, the sequence A–D–A is black–yellow–black (see figure 8.7). But what is that supposed to mean? Only when we look at a yellow swallowtail and remember that Nabokov is a skilled and published lepidopterist do we get his private joke (see figure 8.8).

Nabokov's writings appeal unusually to the senses. For example, he writes about the scent of fresh violets clinging to a woman's black veil or refers to his own father as hard white and gold. It is impossible to say whether he is being synesthetic or poetic. According to his son Dimitri, "he himself realized that imagery is colored for poetic reasons rather than

Figure 8.7
Nabokov's grapheme colors for "Ada" are black–yellow–black.

Figure 8.8
Nabokov's private butterfly joke: the black–yellow–black pattern mimics that of a yellow swallowtail.

instinctive ones. So there is a kind of delicate balance between the two."[25]

Vladimir Nabokov wrote a poem about synesthesia in 1918, when he was only nineteen, about summers in the family estates near St. Petersburg:

A gift exists that is unclear to science. / One hears a sound but recollects a hue, / invisible the hands that touch your heartstrings. / Not music the reverberations that ensue within; they are of light. / Sounds that are colored, an enigmatic sonnet was addressed to you / that scintillate like an iridescent poem by Arthur Rimbaud, their land's conniving crony. / Besides that, there are colors that have sound. / On limpid, melancholy days in autumn upon the purple of a maple leaf / I seem to hear the tremulous and distant hollow re-echo of a horn. / The beauty fades, transformed to simple tunes / a crystal ringing in dahlia's fiery facets, I perceive, / on dry grass midst the cobwebs' motley weave.

Perhaps the most common approach to so-called synesthetic art is for trained individuals to use their synesthesia as a springboard or source of inspiration. In chapter 4 (p. 101) we mentioned how Wassily Kandinsky gradually intellectualized his synesthesia, eventually equating sound and color with Schoenberg's twelve-tone music and seeking a universal translation scheme among the senses. In the latter attempt he was unfortunately misguided because we now know synesthetic couplings are idosyncratic.

A better example of synesthesia as creative springboard is Carol Steen, who incorporates her synesthetic photisms in numerous paintings and sculptures:

These brilliantly colored and kinetic visions, or photisms ... are immediate and vivid.... All my life I had unknowingly created my artwork using this internal landscape, but once I finally discovered what synesthesia was I made the conscious decision to use what I saw.

When I first began creating with my synesthetically perceived, moving, luminous shapes, I was concerned with accurately isolating and depicting them and their source. Now, I work using just one "sense trigger," such as sound ... listening to only one selection of music at a time, played over and over again until the painting or sculpture is finished. A work need not be completed in one day provided I listen exactly to the same music when I return to work ...

Once, unfamiliar music made Carol see a string of shapes appearing from left to right as if on a screen:

The shapes were so exquisite, so simple, so pure and so beautiful that I wanted to be able to capture them somehow, but they were moving too quickly, and I could not remember them all. It is a pity, because I saw a year's worth of sculpture in a few moments.

Carol has four kinds of synesthesia: [graphemes/words → color, shape] + [sound → color, shape, movement, location] + [touch/pain → color, shape, movement, location] + [smell → color]. What she experiences are "intensely brilliant, luminous colors and simple, soft-edged shapes" that are textured. They move against equally saturated color backgrounds, most commonly a black with a "soft inviting texture" like velvet or "grass in the breeze." Her photisms swirl or fall "like the waves of the aurora borealis," then abruptly vanish. Some photisms she has seen only once, whereas others occur frequently enough that she can remember them sufficiently to use artistically. Usually, "they speed by my eyes too quickly for perfect total recall."

Whether or not she paints or sculpts a photism depends on whether she finds it "visually interesting," surely a universal criterion for any artist. There are certain families of colors and shapes that particularly grip Carol. Already in her act of deciding what to depict there is artistic discernment. As she puts it, "some visions are just not interesting or beautiful." It is therefore inaccurate to say she merely reproduces her synesthesia given the fair amount of judgment exerted by virtue of her artistic training and aesthetic temperament. For example, "In creating my sculpture I simplify and choose from the many shapes I see." Usually, she paints when she wants to explore a photism's color but sculpts when interested in using the combinations of a photism's moving shapes.

It took Carol a long time to realize that the colors she saw were those of light rather than pigment. Most of the time, however, highly saturated paints do provide a "faithful mapping" of her color experience, although the best color matches result with "media that rely on light transmission—film, video, computer monitors, blown or stained glass with light shining through, translucent gemstones strung on platinum filaments."

As for shape, Carol was astonished to see that Klüver's form constants already represented many of the configurations she sees: "circles, broken lines, parallel straight lines, parallel curved lines, and zigzags." She does not see lattice or grid forms but does experience "almost geometric shapes that approximate spheres, circles, pyramids, triangles, and squares—but no cubes."

An example of grapheme-based synesthetic art is Carol's sculpture called *Cyto* (see figure 8.9). Made of bronze with a cerulean blue patina, it is based on the colors of Dr. Cytowic's last name. The shapes are stacked vertically on top of one another because that is how she usually senses her moving colored forms:

In this instance they twist, turn, and combine, moving from the tripod base to the two slender forms at the very top. They dance as the shapes danced in my vision. *Cyto* is divided into two almost separate parts, which can seem to occur when I see one set of shapes being replaced by another set.

Cyto is unusual because Carol rarely uses graphemes to create synesthetic art. Grapheme colors are static "and hardly ever do something unexpected," whereas photisms triggered by sound and touch typically do. The latter two are also "quite powerful." Over time she has come to favor acupuncture-induced photisms over sound-induced ones, mainly because there is nothing to do during treatment sessions but "observe them in total peace and concentration," usually with eyes closed. By contrast, sound-

Figure 8.9
Carol Steen's *Cyto* sculpture, bronze and steel with blue patination, conveys the shapes, color, and twisting movement of the first two syllables of Dr. Cytowic's name.

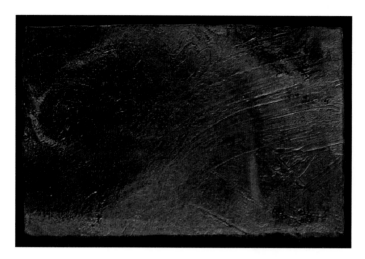

Figure 8.10
"Red Sharp" depicts a single shape streaking across a golden photism, instantly seen by Carol Steen when she got a tetanus shot.

induced photisms are less brilliant, perhaps because painting requires eyes open in a brightly lit studio. The latter also causes the problem of divided attention while painting to music.

Bright chrome orange is typically her color of pain, though the depth of the acupuncture needle affects what she sees. Most often, photisms occur when the needle is either being inserted or removed. Touch-based artistic compositions tend to be simpler but stronger because Carol chooses to depict a single photism shape from her overall experience. The painting titled "Red Sharp" (see figure 8.10) is such an example.

Voice is another reliable synesthetic trigger because some people's voices cause "wonderful colors," even though sound-induced hues are less saturated and brilliant than their touch-induced counterparts. However, an unusual experience once occurred in near-total darkness on a moonless night at Wolf Howl Pond in Canada. Knowing that wolves howl back to a human imitation, Carol's friend tilted his head and gave "a long, low howl in the most perfect green, finishing on a note of red as his pitch dropped toward the end of the call." The painting in figure 8.11 depicts what she saw while listening to him.

Lastly, there is the synesthetic artist inhabiting the moment such as Marcia Smilack. Her multiple synesthesia types are: [graphemes → color,

Figure 8.11
"Michael at Wolf Howl Pond" is what Carol Steen's friend's green and red voice looked like as he howled like a wolf on a dark moonless night.

gender, personality) + [sound → shape, color, motion] + [color/shape → sound, motion, texture] + [touch → color, shape] + [sequences → number forms]. Marcia calls herself a reflectionist given that she photographs reflections in water "whenever I hear a chord of color . . . to what I see." For example, "Cello Music" (see figure 8.12) captures the reflection of a setting sun whose "golden hue, together with the wave formation, creates the sound of a cello," whereas in "Red Buoy 1" (see figure 8.13) it was a bright red sound, loud and irresistible like a siren, that captured her attention.

Marcia has no training in either photography or art, so her photographs are never "posed." Rather, something—a color, a shape, a motion—catches the corner of her eye and prods her to swing the camera round to capture the moment. She quickly responds intuitively to novel stimuli rather than thinking about composition. Understandably, her technique yields a lot of hit-or-miss images. While some can be arresting, there is neither the intention nor the possibility for audiences to understand the synesthetic

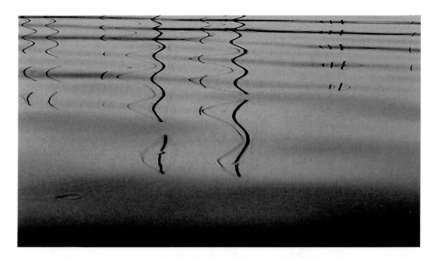

Figure 8.12
"Cello Music." For Marcia Smilack, the golden brown and the wave motion create the sound of a cello.

Figure 8.13
"Red Buoy 1." For Marcia Smilack, bright sun reflecting off a red buoy produced a loud siren-like sound—not a loud annoying siren but "like the sirens in Ulysses which neither he nor I could resist."

experience that leads to them. Thus, they must stand alone on their own merits.

David Hockney

The British painter David Hockney (b. 1937) has [sound → color, shape, movement] synesthesia. It did not, of course, contribute to the paintings that made him famous given that these were "silent" works. Only later when he designed costume and stage sets for the Ballet de Marseille and the Glyndebourne and Metropolitan Operas did a new element appear in Hockney's work. Unlike his earlier pictures, he explicitly conceived his designs *to the music*. Certain comments by the artist made Dr. Cytowic suspect he was synesthetic:

I find that visual equivalents for music reveal themselves. In Ravel, certain passages seem to me all blue and green, and certain shapes begin to suggest themselves almost naturally. It's the music that attracts me to doing the set designs rather than the plot.

Dr. Cytowic's suspicion was later confirmed by examination and experiments conducted in Los Angeles (September 11–12, 1981), following receipt of a letter in which Hockney wrote:

I know it seems a long time to take to answer your interesting letter, but I have carried it out with me for a few months sometimes thinking of replying and then putting it off, then thinking I put it off for a good reason. Would it tell me any-thing—or do I really want to know, etc.

I must admit my first reaction to it was that you were trying to describe academi-cally something I'd always thought and explained away as "poetic." I'd never heard of synesthesia.

Anyway here I am replying to your note. Curiosity has got the better of me and so perhaps we could arrange a meeting.[26]

Unable to read music, Hockney listens to it over and over. "I'll listen to the specific music constantly while I'm working," he says. The music guides the actual movement of his arm. In painting Ravel's *L'Enfant et les Sortilèges* for the Metropolitan Opera in 1981, the "musical description of the tree in the garden has actual weight, like a tree has. I drew the form of the tree to the music. I painted the sets to *Rossignol* the same way—to the music."

Like Messiaen, Hockney is an example of an innate talent within synesthesia that is modified by personal intellect, creativity, and artistic vision. Above all, he is that rare artist articulate enough to convey to the rest of us a sense of how he translates his private synesthetic experience into moving public artwork.

Hockney was not exposed to music as a child and has no musical talent. "It wasn't until I was forced to do something about it in 1974 with *The Rake's Progress*." Apprehensive about having to conceive a visual piece for the opera score, he gave up trying to analyze the music only to realize he experienced something else. "But then it was largely *involuntary* and I do 'get it,' something clicks and all of a sudden you hear and feel more of the music." In the Ravel, the tree music (of *L'Enfant et les Sortilèges*) dictated its volume and weight. By this, Hockney means a volume and visual area that corresponded to the physical shape of the music. He would actually paint with a long three-foot brush, articulating at the shoulder rather than the fingers and wrist, while he listened, the music dictating his arm motions—the lines, curves, dots, and blots as well as the color and overall dimensions. "In all operas I've done, the music gives me the set—the color and the shape. In the [Stravinsky] *Oedipus Rex* there was not much color but lines and sharp things that suggested cross hatchings." This aptly describes the category of grid form constants.

A pilot study performed in Los Angeles confirmed the existence of absolute and relative effects in a sound–color matching task similar to the one described in chapter 4. The forced-choice test used 120 trials for each stimulus. Instead of verbal labels, however, in this study Hockney chose from a complete set of color chips. In addition to the effect of single tones on color matching, further experiments examined melody via major and minor arpeggios and triad chords. Chords are more like tones than arpeggios, whereas arpeggios are ascending tones strung together. Tones that Hockney perceived as high evoked reds, pinks, and yellows, whereas minor arpeggios yielded a very restricted response set of blues and purples. For Hockney, the thing that most predicts a restricted response is melody. Thus, the pilot studies confirmed what the artist himself acknowledged—that the music itself, its melody, dictates shape, color, and movement.

Once Hockney's attention became focused as a result of painting stage sets to music, he occupied himself with color and space in a new way. He says he wants to construct a space in which one can see around corners as distinct from the regimented single-point perspective that Canaletto, for instance, practiced. For Hockney, manipulation of space is possible through the use not only of pigment but also of colored lights. Over the years, Hockney has experimented with increasingly complicated colored lighting systems. The best way to reveal his thoughts on color, light, and form is to let his conversation with Richard speak for itself:

REC: You've taken some joking about your "light box," which is actually a scale model of the Metropolitan Opera stage, complete with a colored lighting control system. You say you alter your design sketches while viewing them in the light box. Can you explain its importance?

DH: Not many people use color—real color—in the theater. If you're going to use color, then you have to have colored lights; otherwise you'll never know what color to paint. You have to test it. I had that box made in London when we were doing the Ravel because of what happened when we were doing [Satie's] *Parade.*

When I finished the drops for *Parade* last year, John Dexter said, "That's nice, we'll just put white lights on them." I said, "No. You put red and blue lights on them because that's what will make it magic. That's what will make it sing." It took them some time to realize that I was right. In London, five months before we staged it, I lit it crudely, and then slowly we devised the scale model machine so that I could time the color changes to fit the musical changes.

REC: Do you always work with the lights on your sketches?

DH: Constantly. I keep fiddling. Looking, listening, and playing with the lights, and it simply takes you a long time because you keep hearing more and more in the music.

REC: During the matching test we did yesterday, you said that the "correct" color stood out when the music was on. How so?

DH: There's a shimmer so that one of the colors stands out at the moment that the music is on. When the sound is coming, there's an extra vibrancy to the color.

REC: A eureka sensation?

DH: Yes, an intuition that says, "This is it!"

REC: How can color be used to control your sense of space?

DH: Blue has this quality of being spatial, which other colors do not. The more of it there is, the more you feel of it. I did the same thing with light and dark in *Oedipus Rex*. [At the end of that opera, Hockney projects gold light out onto the proscenium and the front of the house to incorporate the audience into the opera. Hockney's hand outlines a cross in the air as he speaks.] The music is like this—horizontal and vertical, very geometric. I projected the gold onto the side of the proscenium to give it incredible weight, and to make it big. Boom, boom. The first quality of *Rossignol* that caught my ear was of transparency. I listened first. I don't look at the score because I can't read music. When I first heard it, it doesn't occur to you that it's Chinese. What one hears is transparency. It's about transparent things, night, moonlight, water.

REC: Your set for *Rossignol* is also all blue, it's monochromatic.

DH: But there are a lot of different blues in it. It's not just all one blue.

REC: How does the blueness fit in with the finished work at art?

DH: The blueness and this sense of transparency that's in the music made me think of the very refined beautiful china of the 17th century. Not the overdone 19th century stuff with dragons, but much simpler, purer versions. I went to the Victoria and Albert Museum in London and in just two cases of pottery I took about 150 photographs and that's where the trees, mountains, and people came from in the set. Since the three Stravinsky operas are all in some way ritualistic, each piece is united by a circle motif. John Dexter wanted a disc on the floor, but I wanted a transparent blue circle. So I made it a blue china plate.

REC: What about revisions? How much of the color and shape comes to you from just hearing the music and how much do you bring to it by "intellectualizing" or intentionally revising?

DH: It's like the shimmer with the colored chips. I know visually when the color or the lines fit the music. We made about 10 palace drops for *Rossignol*, and I thought, "It doesn't fit the music." The lines weren't right. Each time you listen, though, you hear more and more in the music. In the end, it looks like Chinese cubism to me, but it fits the music. It looks three dimensional, too, because I've painted it on black velvet. But it's completely flat. Black has an enormous space to it. Once you grasp the illusion of three dimensions, you don't scrutinize it.

We went through 27 versions of *The Rite of Spring* sets before they fit the music too. Most of the problem there, however, was getting the color rather than the lines right.

REC: You say that you actually paint while the music is playing.

DH: It's very hard to describe because it's ineffable. With Ravel's tree music, for example, I remember drawing the lines of the tree to the music because it had a weight. You know how a tree has volume and a weight? I drew it during the music.

For *Rossignol* and the others it was the same. When I'm working on it I will play with music constantly. Normally when I work I do not work to music. I don't like it as background because you find you either listen or you don't. It could be trashy music—a little ballet or something, *Swan Lake*—where you wouldn't be too distracted. But I couldn't possibly work and listen to a Beethoven quartet. I couldn't because I would lose the lines of what I was drawing at the moment.

REC: Does the actual performance, the singing and musicians, influence the way the set looks to you?

DH: Yes, it works both ways. One has to be flexible, too, although I'm quite insistent on some points, particularly the color. There are many things to consider. You work with a director, and people have to move in your sets. John Dexter [the director] tends to make diagrams, that's his style. I said, "OK, but I'll make pictures of them since I like pictorial stuff." Then you're told the chorus has to be there in the middle—36 of them—or they have to be there from the beginning. Well, that's ruined my picture, I think. Then I think, well, there's no need to if you stick them in the middle right from the start, then you'll forget about them. They'll disappear. Then the musical people tell you there must be something behind the people; otherwise you won't hear the sound, and you know if it doesn't sound right, it's going to look hideous. And so there's no point arguing. It's very complicated.

Deliberate (Nonsynesthetic) Contrivances

There is some confusion over the word "synesthesia" being used throughout its long history to describe wildly different things: from poetry and colored-light music to deliberately contrived mixed-media applications such as Laserama, son et lumière, and smellavision. In other words, we must take care to separate pseudo-synesthetes such as Georgia O'Keeffe or Alexander Scriabin, who intellectually applied ideas of sensory correspondence to their art, from artists who truly had the perceptual variety such as Wassily Kandinsky and Olivier Messiaen.

It will not do to claim that someone "might have been" synesthetic just because they wrote colored music or titled their paintings with musical terms. For example, Georgia O'Keeffe's "Music—Pink and Blue II" (1919) and "Blue and Green Music" (1921) are impressionistic canvases, whereas Roy De Maistre's "Rhythmic Composition in Yellow–Green Minor" (1919) is based on a premeditated "colour harmonizing chart" that deliberately pairs up pitch with color. There is nothing in either artist's history to suggest they ever experienced the involuntary, consistent, and lifelong photisms characteristic of genuine synesthetes.

Likewise, Paul Klee famously said, "One day I must be able to improvise freely on the keyboard of colors: the row of watercolors in my paint box."[27] However, Klee was an accomplished musician, who just as famously explored the similarities between auditory and visual rhythms at the Bauhaus, using the term "polyphonic," meaning many-voiced, in his painting titles to convey "sounds" he created by layering color, line, and shape on one another. "The simultaneity of distinct themes is not something that is unique to music," he said. Despite his close absorption in the problem of sound–sight relations, however, Klee was not synesthetic. Rather, his explorations were clearly academic as laid out in Hajo Düchting's book about Klee, *Painting Music*. It detracts nothing from the drawings' intrinsic interest or beauty to make the distinction.

Scientifically, we need positive evidence that a given artist is synesthetic, which is lacking in most of the usual suspects from the late nineteenth and early twentieth centuries. Kevin Dann, who explores this epoch in *Bright Colors, Falsely Seen: Synesthesia and the Search for Transcendental Knowledge*, convincingly shows that a cultural, somewhat mystical interest in sensory fusion permeates most of these false claims.[28]

Many individuals often suspect Charles Baudelaire (1821–1867) of being synesthetic because in 1857 he asserted that "sounds are clad in color." In his famous poem "Correspondences," he further claims *Les parfums, les couleurs et les sons se réspondent*. It is conceivable that Baudelaire's well-known use of hashish rendered him temporarily synesthetic, but no other shred of evidence exists that he had perceptual synesthesia as we understand it today. A bit later, in 1871, another French poet, Arthur Rimbaud (1854–1891), linked color with vowel graphemes in *Le sonnet des voyelles*:[29]

A black, *E* white, *I* red, *U* green, *O* blue;
Some day I'll crack your nascent origins

A, Black hairy corset of gaudy flies
That bumble around cruel stinks,
Gulf of darkness

E, Canyons of vapors and intense spears of proud glaciers
White kings
Flutter of flower clusters

I, Deep reds. Coughed up blood
Laughter of lips, beautiful of anger, of petulant ecstasy

U, Cycles, Divine vibrations of seas growing green
Piece of the pasture sown with animals
Piece of the wrinkles that alchemy stands on studious brows

O, Supreme clarion full of strange silences
Silences crossed by worlds and angels
O, omega. That violent ray of his eyes

Rimbaud later claimed to have invented vowel colors—literally, *"J'inventais la couleur des voyelles!"*—but nowhere in his correspondence or biographies do we encounter sentiments typical of grapheme → color synesthetes. He does not even follow the Berlin and Kay color typology wherein *A* is commonly red, *I* and *O* white, and so on for both synesthetes and non-synesthetes.[30] Rather, the poem's language belies Rimbaud's arbitrary and imagistic color associations.

A much earlier poet, Basho Matsuo (1644–1694), is considered Japan's finest practitioner of haiku. In one well-known poem he combines three perceptions of sound, scent, and twilight, juxtaposing the verb *tsuku*, meaning "to strike," with "smell of blossoms" in such a way as to connote that the flowers ring their scent into the fading twilight:

Kane kiete	As the **bell tone** fades
hana no ka wa **tsuku**	**Blossom scents** take up the **ringing**
yube kana	Evening **shade**

Compared to the immediacy of synesthesia, the poem's gradual transition from sense to sense indicates that Basho is being metaphoric rather than synesthetic. Given the regular and lawful associations that do exist among different sensory dimensions, however (refer back to figures 4.4 and

8.1), we may rightfully ponder whether Basho is being strictly poetic or making explicit an inherent, but normally implicit, interpenetration of sensation and action. Regardless, he is not synesthetic as we understand the condition's expression.

Composers Alexander László (1895–1970) and Alexander Scriabin (1871–1915) both wrote color–light music, and others followed. László never claimed synesthetic experience, whereas Scriabin spoke of sound–color correspondences and claimed musical keys instilled an overall sense of color. His case was reported in 1914 by the British psychologist Charles Myers,[31] a fact that tells us more about the state of psychology at the time as well as the synesthesia euphoria[32] that gripped academic minds of the period (refer back to figure 1.3). We observe that Scriabin assigned colors to keys rather than note pitches as individuals with colored hearing typically do. Furthermore, we find his juxtaposition of color and key far too neat: Scriabin's colors are systematically arranged in their rainbow order—red, orange, yellow, green, blue, violet, purple—while being neatly paired with the orderly modulation of keys that musicians call the circle of fifths (see figure 8.14).

On casual inspection, Scriabin's affective and symbolic tone–color associations (e.g., F minor is blue, "The color of reason," F major "The blood red of Hell") appear to be directly quoted from Madame Blavatsky's color–note table as described in her Theosophist tract *The Secret Doctrine*.[33] Here again the zeitgeist easily comes through. In 1911 Scriabin composed *Prometheus, The Poem of Fire* for large orchestra and piano, with organ, choir, and *clavier a lumiéres*. At the time, the *clavier a lumiéres* was nonexistent, conceived of as a mute keyboard that could control the play of colored light in the form of beams, clouds, and other effects flooding the concert hall and culminating in a white light so strong as to be "painful to the eyes." In the score its part, marked "luce," is written in conventional musical notation on the uppermost staff (see figure 8.15). Technical reviews appear in period volumes of *Scientific American*.[34]

Advancements in color film and recorded sound from the early 1930s onward allowed exiled German artists like Oskar Fischinger (1900–1967) to escape the limitations of static painting and set abstraction in motion. Hand painting film frames enabled elaborate sequences of geometric shapes that transformed in time to a synchronized soundtrack. Whereas Fischinger[35] claimed to aim "only for the *highest* ideals—not thinking in

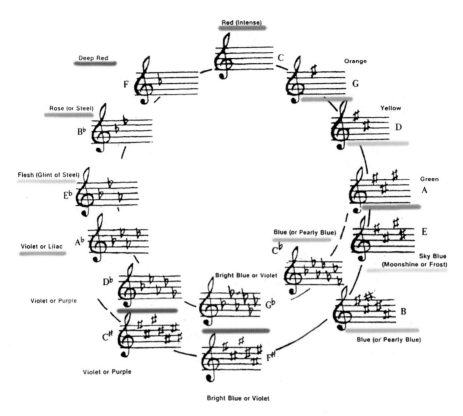

Figure 8.14

Scriabin's rainbow progression of colors pairs up with the clockwise arrangement of the twelve keys in an order of ascending fifths, showing that after twelve such steps the initial key is reached again. His system is a deliberate contrivance rather than a manifestation of perceptual synesthesia.

terms of money or . . . to please the masses," Walt Disney,[36] for whom Fischinger eventually worked, said, "Everything that has been done in the past on this kind of stuff has been cubes and different shapes moving around to music. . . . If we can go a little further here . . . the thing will be a great hit."

Disney's optimism resulted in *Fantasia* (1940), to which Fischinger contributed the animated opening of Bach's Toccata and Fugue in D Minor. He later quit, citing artistic differences and conditions he felt stifled his creativity. Despite *Fantasia's* poor 1940 box office, history judged the film a classic. *Fantasia's* history traces back to those abstract artists for whom

Figure 8.15

(*Left*) Title page of Scriabin's "Prometheus." Notation for the light instrument is on the topmost staff, marked "luce." (*Right*) The cover of the score, to which Scriabin attached enormous importance, shows a flaming sun with an androgynous face enclosed in a world lyre and surrounded by magical and cosmic symbols—stars, comets, and spiraling clouds. Colored orange—"the color of fire"—and drawn by the composer's Theosophist friend Jean Delville, it reflects Scriabin's interest in mysticism.

Figure 8.15
(*Continued*)

music was the highest art (and which they thought painting should emulate), to color organists of earlier centuries to the present, to the links between color and emotion formulated by the Symbolists and like-minded artists, to experimental filmmakers, and, more recently, psychedelia, Laserama, and digital media. While none of these achievements is synesthesia in any strict sense, all of them draw on understanding that there are equivalent associations among different dimensions of sensation that we best understand as metaphor.

At the opening night of the Hirshhorn Museum's 2005 exhibit of "Visual Music," at which Richard later gave a galley talk, a curator went on about a machine whose moving pendulums directly inscribed a soundtrack. "Here you have the purest example of a movement-to-sound synesthesia," he enthused. To which Richard countered, "No, you're not even wrong! Mechanical transcripts are not synesthesia because machines cannot perceive. Only a human brain, necessarily embodied, can experience synesthesia. Moreover, a machine could never relate what that experience is."

Synesthesia and Creativity

We have already laid out several key features linking synesthesia with creativity throughout this book, so our argument should now be easy to follow.

We showed earlier that metaphors are not arbitrary but follow rules. Like synesthesia, they too are *directional*, going, for example, from hearing to vision ("loud tie") much more often than they travel in the other direction ("bright note," but never "red sound"). That is, "A is B," but not "B is A." We gather from these facts that anatomical limitations allow some types of cross-activation but not others.

Another reason for directional asymmetry may have to do with the fact that synesthesia is meaningful as well as perceptual, and that meaning itself is directional. Consider, for example, that in grammar subject–verb–object relationships are not reversible or that metaphoric references are also asymmetrical—"man is a beast," but not "beast is a man."

From a genetic standpoint we raise two issues: the gene's dose and its scope of expression. In synesthesia, an optimal amount of gene is expressed—too much and it becomes bidirectional or manyfold, leaving a minority of individuals confused or overwhelmed. Because synesthesia is really not good for anything, 1 in 23 individuals carry a useless trait from a genetic point of view. But like sickle cell anemia,[37] we think the trait has hidden importance, which is why it is maintained so commonly in the population. We cannot definitively say what its advantage is yet but venture a guess that it has to do with creativity—specifically, an ease for making metaphoric cross-connections.

For example, psychologist Harry Hunt in Ontario points out[38] that some children are described as fantasy prone wherein in adults we speak of the personality trait of high absorption. The latter is characterized by a developed imagination and a greater openness to metaphor.[39] Hunt suggests that "so introspectively sensitized" a person has "a greater capacity for immersion in the immediate cross modal properties of our ongoing consciousness, making its synesthetic features more noticeable." Once the synesthesia gene is identified, and if it is found to be also expressed in nonsensory brain areas, it could be that rather than being specific to synesthesia it more generally codes for absorption and openness to experience, hallmarks of creative individuals.

We mentioned the impression that synesthesia appeared to be more common among artists, composers, and novelists. It is just such creative types who are adept at making metaphors, seeing connections among seemingly unrelated things. We already know that high-level concepts such as numbers, faces, and pitches are represented in brain maps the same way that percepts are. Thus, there is no reason to believe that other concepts are not similarly represented. The idea is that synesthetes have more abundant connections among conceptual maps and accordingly are more facile at linking superficially unrelated concepts and seeing deep similarities in them. Objectively, synesthetes are in fact facile at fathoming deep similarities in seemingly unrelated realms.[40] Likewise, they have a higher aptitude than nonsynesthetes for figurative speech,[41] though this is likely an effect rather than a cause.

The angular gyrus at the junction of the temporal, parietal, and occipital (TPO) cortices is known to be crucial for forming cross-concepts because it is a functional crossroad of feeling (parietal), hearing (temporal), and seeing (occipital). The convergence of three sense modalities results in abstract, modality-free representations about objects in the world, such as the concept of jaggedness or sharpness in the "bouba"–"kiki" illustration. Interestingly, Ramachandran[42] has done the "bouba"–"kiki" experiment on individuals with left angular gyrus lesions and finds that they cannot make shape–sound abstractions even though they are normal in all other respects. Such individuals also cannot explain the meaning of proverbs (e.g., "Don't cry over spilt milk"), whose understanding relies on making abstract cross-connections. We have already seen in discussing number forms in chapter 5 that the *left* angular gyrus also concerns itself with sequence and ordinality, but it may be that the *right* angular gyrus is required for their sense of spatial configuration because clinically that is what the right parietal region does. Accordingly, Ramachandran and Hubbard suggested that the left TPO junction may be a network node for synesthetic cross-sensory metaphors ("loud tie") whereas the right TPO may be a node for spatially characterized orientational metaphors ("*declining* health made him *step down*").[43] There is no reason why the TPO junctions could not also support ontological metaphors, although there is currently no evidence addressing this. Neuroscience has simply not studied metaphor comprehension in depth, but these ideas constitute testable hypotheses.

Comparing human brains to those of apes and other mammals, one finds a huge development of the angular gyri and TPO junctions. Earlier we proposed a cognitive continuum of perception → synesthesia → metaphor → language. Let us now flesh out how such a continuum might come to exist.

One of the earliest behavioral neurologists to call attention to structures at the TPO junction (such as the angular gyrus and inferior parietal lobule) and their role in making stable cross-sensory associations was Norman Geschwind. He pointed out the following in 1964:[44]

The ability to acquire speech has as a prerequisite the ability to form cross modal associations. In sub-human forms, the only readily established sensory–sensory associations are those between a non-limbic (i.e., visual, tactile, or auditory) stimulus and a limbic stimulus. It is only in man that associations between two *nonlimbic* stimuli are readily formed and it is this ability which underlies the learning of names.

That is, an ape can make the emotional (limbic) association that a banana is positively valenced (it tastes good). It can make cross-modal transfers between touch, vision, and hearing, and the limbic inputs of associated experiences of hunger, thirst, fear, sex, pleasure, disgust, and so forth, whereas humans can further give the banana a name and associate other nonlimbic qualia with it (yellowness, slipperiness, and eventually banana-peel jokes). Even young children readily make cross-modal associations, recognizing as identical by age four an object seen alone and then felt in the dark.[45] The inferior parietal lobule, just below the angular gyrus, figures prominently in language and receives inputs only from association cortices. Developmentally, it is late to insulate with myelin, is late to mature architecturally, and is one of the last areas in which dendrites appear.[46] It is precisely here where nonlimbic intersensory cross-connections are powerful.

Other animals make cross-sensory associations. The vocal–kinesthetic mirroring of parrots is probably executed by the cerebellum, whereas the cross-sensory translations of dolphins who learn visual-token sign language require the forebrain.[47] Both animals, along with the great apes, lack the three-way cross-sensory couplings among vision, hearing, and gesture that support the human capacity for semantic abstractions.

Geschwind was explicit in discrediting the idea that "verbal mediation" is the means by which humans achieve cross-sensory transfers:

> . . . . it cannot be argued that the ability to form cross-modal associations depends on already having speech; rather we must say that the ability to acquire speech has as a prerequisite the ability to form cross-modal associations.[48]

In other words, the ability to form intersensory associations allows the subsequent development of speech, which frees us from the tyranny of the senses[49] and the immediate pleasure–pain principle of the limbic system. Once speech is possible, humans can make other intermodal associations, which form the higher and most abstract kinds of cognition.

One of Geschwind's main points is that language depends on stable intersensory associations, especially auditory–visual and auditory–tactile. These are, unsurprisingly, common modes of synesthesia. In our view, cross-modal metaphor is an abstract linguistic derivative of synesthetic experience, thus the continuum going from perception → synesthesia → metaphor → language.

The detailed genetic expression of synesthesia might determine whether an individual will have one, two, three, or more kinds of synesthesia. Having different kinds of synesthesia may mean that the gene(s) are expressed in one to several different brain areas. We already stressed that synesthesia can be either perceptual or conceptual. Broader gene expression may increase the likelihood of more-than-usual connections between conceptual maps, and thus a higher level of creativity.

The evidence for higher creativity in synesthesia is not just anecdotal or obtained through the survey methods reported above. For example, testing fine art students[50] showed that synesthetes scored significantly higher than controls on four experimental measures of creativity. Better objective measures by Catherine Mulvenna, Dana Sanders, and Edward Hubbard show that synesthetes score higher than age-, gender-, and intelligence-matched controls on three measures of creativity: fluency, flexibility, and originality. Collectively, the team confirmed four synesthetes at the University of Glasgow and five at the University of California, San Diego, and tested them using measures such as Raven's progressive matrices[51] and the Torrance Test of Creative Thinking.[52] On the first, a test of intelligence, synesthetes scored as a group significantly higher than controls, demonstrating somewhat stronger ability at spatial and orientational metaphors. The second, a creativity test, rates fluency, flexibility, and originality of responses. Here again, synesthetes scored significantly higher in all three measures of creativity. Thus, preliminary evidence supports the supposition that synesthetes are more intelligent, more abstract, and more creative than their nonsynesthetic counterparts. Synesthetes also show higher aesthetic sensitivity compared to the rest of the population.[53]

9 Inside a Synesthete's Brain

Synesthesia does not localize to any one spot in the brain. Rather, we need to think in terms of neural networks that connect multiple brain regions because synesthesia seems to be the perceptual result of *increased cross talk* between brain areas.

How might such an increased cross talk come about? Is it from a failure to prune connections at an early age? Or is cross-wiring present in everyone, synesthetes and nonsynesthetes alike, with only synesthetes having the pleasure of experiencing activity above the threshold for conscious awareness? To understand the brain of a synesthete, let us first turn to how the normal brain is organized.

A Division of Labor in the Brain

To the naked eye, the human brain looks the same everywhere as we glance over its ridges and furrows. Upon closer examination, however, we find that different types of information are processed in different parts. Specific brain areas are involved in hearing, speech, memory, movement, emotion, and so forth (see figure 9.1a). When we subsequently zoom in on a single sensory area such as vision, we discover that separate populations of cells are sensitive to different aspects of sight, such as motion perception, edge detection, facial recognition, color perception, and so on (see figure 9.1b). Despite surface appearances, therefore, brain territory is not uniform at all, but instead highly specialized at all levels.

As the details of neuroscience unfold, we find ourselves looking at a strange commune of neural subsystems that care about smell, color, touch, hunger, pain, goal setting, prediction, and countless other tasks. The networks supporting these various functions continually operate all at once,

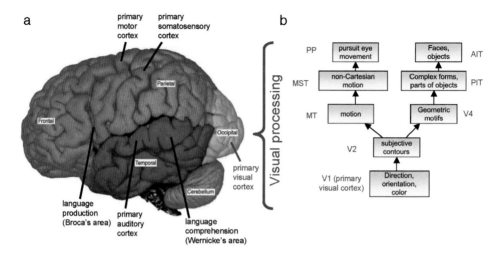

Figure 9.1

Division of labor in the brain. (*a*) The brain is roughly divisible into areas specialized for movement, sensing the body parts and internal viscera, various aspects of language, vision, hearing, and so on. (*b*) Many subspecializations exist within the sense of vision. As we move farther away from primary visual cortex (V1) into higher visual areas, we encounter brain areas specialized for increasingly more complex stimuli, such as detecting motion, houses, or human faces. The box diagram summarizes the vast array of data processing undertaken by populations of cells within the larger visual cortex. V1, primary visual cortex; V2, secondary visual cortex; MT, medial temporal; MST, medial superior temporal; PP, posterior parietal; PIT, posterior inferotemporal cortex; AIT, anterior inferotemporal cortex.

interacting with astonishingly useful results. For example, the auditory system deals with air compression waves while the visual system analyzes patterns of photons. The startling thing about brain function is that all these disparate systems find ways to work seamlessly together, even to help one another. When trying to understand a new object, for example, you might look at it, turn it over in your hands, shake it, smell it, and so forth—combining different senses to wind up with an integrated impression of the object. The challenge for brains is to allow disparate regions to perform different tasks while also working together to share information. For synesthesia researchers, the question comes down to how and why this balance among networks is different in synesthetic brains.

To understand the synesthetic brain, we must first ask how the division of labor in normal brains comes about. From an evolutionary standpoint, human brains need to answer the same question that governments and organizations do: how specialized is it useful to be? Too much division of labor leads to a breakdown in communication, whereas too little reduces efficiency and precludes expertise. Brains manage to strike a balance. On one hand, they must distinguish between incoming signals such as sound and sight (say, a growl vs. a pineapple). On the other hand, brain areas cannot speak totally different languages, because a primary survival task is putting all the signals together to accurately interpret the world. This is how brains can learn that the growl goes with the sight of the lion, whereas the pineapple image goes with the sweet smell and taste.

Thus, brains must split the difference between too much specialization and too little, something akin to the debate over federalist versus states' rights in the Colonial Congress of the United States. As we will see, the phenomenon of synesthesia has caused neuroscience to start paying attention to the amount of integration that is useful for brains. Too much integration means that signals cannot be differentiated. Too little integration makes it impossible to extract the underlying similarities between different sensory domains.[1,2]

For example, in coming to a decision about what one just saw, what to do next, or what to do later in the day, a range of signals must be brought together fruitfully. This is a tough problem for brains because relevant factors in the calculation include the state of your inner milieu (are you hungry, hot, or tired?) as well as memories, knowledge of your own skills and capacities, and of course diverse sensory input. Each of these in turn involves various integrating steps. For example, gazing at a swimming fish, we effortlessly see its shape and motion as inseparable parts of a single object. Such integration is often crucial to predictions and motor choices, such as catching the fish if you are hungry or avoiding it if it is a shark. Although you may have heard the brain celebrated for its parallel processing, it should be equally appreciated for its ability to quickly squeeze down all of its parallel processes into a *single* behavioral output. For example, a running animal can go left or right around a tree, but it cannot do both.

Thus, despite the extensive division of labor in the cortex (see figure 9.1b), none of that division is perceptually apparent to us. Instead, we enjoy a unified picture of the world rather than a separate experience for vision, another for hearing, a third for touch, and so on. That is, you perceive an apple tossed at you as a single object, even though it is actually composed of many pieces of information all bound together. Different parts of your brain see that the apple is something red + round + edible + approaching at a certain speed. You see a flying apple. How brains produce a unified impression of the world while dealing with separate processing streams that also operate at different speeds is called "the binding problem." Synesthesia provides an inroad to this puzzle by showing us the flip side of the binding problem—that is, when more than normal is bound together.

How does the brain solve the binding problem? In theory, there are a few possible strategies it might use. First, all the subsystems could employ a common protocol for sharing information. While software developers like this solution, it is not typical of how Mother Nature solves problems.

Another strategy allows all subsystems to respond in their own ways to the available data. In this approach there is a separate mechanism that coordinates the results. For instance, as in the military, a hierarchy can exist in which each level has its own rank of control and can pool data from lower levels. However, the massive feedback found in the mammalian brain does not easily lend itself to a hierarchy model, so this, too, is unlikely.

The most likely alternative for coordinating subsystems is that *feedback connections* force neuronal populations to "come to agreement" regarding their patterns of activity. After all, in our apple example, there is no single anatomical location in the brain where red + round + edible + moving converge from all the different systems to create a flying apple. Instead, specialized areas densely interconnect with one other. The connections are sometimes direct but more often pass through other areas, forming a massively recurrent network of networks. It is *within* this global network that unified percepts emerge. Recurrent connections allow information to simultaneously flow forwards and backwards among areas, resolving conflicts between differently specialized cells responding to the same stimulus. In this way, activity in different subsystems is coordinated without the need for a supervisor.[3] In this framework, the unified image of the world emerges from a network of activated cortical structures. It does not come from a single area at the top of a hierarchy but instead comes from the simultaneous activity of several subsystems—including those involved in sensation, expectation, memory, and cognition. The whole process critically depends on feedback.

The rich tapestry of feedback circuitry dashed the hopes of early neuroscientists that brain areas would be understood as successive stations along a conveyor belt. Instead, it indicates that brain areas depend on the simultaneously interwoven connections they exchange. Neuroscience is currently working to understand these interconnecting dynamic loops. It is clear that drawing conclusions about the function of a brain area from single-cell recordings would be as misleading as studying the global economy based on one individual's credit card statement. In physics, isolating a part of a system allows you to understand it directly; this principle generally does not hold true in biology, and especially not for the networks of the brain.

Cross Talk in the Normal Brain

In chapters 4 and 8 we examined the possibility that everyone is latently synesthetic. By looking at cross-sensory illusions, we established that different sensory channels densely interconnect in the normal brain.

For example, in the ventriloquist illusion, the ears hear sound from one direction while the eyes see a moving mouth in a different location. The brain incorrectly concludes that the sound emanates from the mouth's location because the sight of moving lips influences where we localize the sound. This illustrates the natural interconnectivity between visual and auditory parts of the brain, which collaborate to generate a single, unified perception.[4]

We also mentioned the McGurk illusion, in which the sound of /ba/ is perceived as /da/ when it is coupled with visual lip movement associated with /ga/. Again, the McGurk effect indicates that sight and sound cues rapidly combine at an early processing stage, even before any decisions are made about phoneme or word categories.[5] Studies using fMRI confirm that seen speech influences the perception of heard speech at a very early processing stage.[6] Also supporting the conclusion that sight and sound are quickly coupled is the "auditory driving illusion" in which the steady rate of a flickering light appears faster or slower depending on whether accompanying beeps are presented at a faster or slower rate. The extensive cross talk does not depend on conscious awareness: for example, facial expressions can modify the perceived emotion in the voice of a speaker even when facial expression is not consciously perceived.[7] And it's not only sight and sound that are connected: in an example linking touch and vision, a sudden touch on one hand can temporarily improve your vision near that hand due to feedback from multisensory areas.[8]

Overgrowth and Pruning

The immature brain has substantially more connections between and within areas than the adult brain has. Some of these are removed by pruning, while others remain. In the fetal monkey, for example, about 70%–90% of connections to V4 come from the primary auditory area and from higher areas, whereas in the adult the proportion diminishes to 20%–30%.[9]

If the synesthesia gene(s) led to a failure in pruning these prenatal connections, then projections between V4 and other brain areas would persist into adulthood, leading to color synesthesia in this example. The less pruning, the greater the activity in cross-connected pathways and the greater the likelihood of reaching conscious awareness. In the nonsynesthetic instance in which pathways are heavily pruned, what residual activation remains may be sufficient to establish cross-modal mappings, such as the conceptual relationship between a sharp shape and a sharp sound or high pitches with bright lights—all while remaining insufficient to reach consciousness as synesthesia. Or the cross talk may reach consciousness only under special circumstances, as with brain damage, drug use, or disease.

Carrie Armel and Vilayanur Ramachandran describe just such a case in which an individual progressively lost vision starting in childhood until he became totally blind at age 40.[10] A few years later he experienced touch as spots of light. It required a firmer touch (above the detection threshold) to evoke synesthesia, and the synesthetic thresholds were constant over weeks, indicating the effect was genuine. A likely explanation is that visual deprivation allows tactile input to activate visual areas by letting existing connections between the areas become hyperactive. The connectivity does not depend on the growth of new connections, as evidenced by the rapid changes in experiments featuring blind or blindfolded subjects.[11] For example, as newly blind individuals learn Braille, the brain area corresponding to the reading finger expands as unused V1 changes its functional assignment from sight to the feeling and "reading" of Braille.[12] More to the point, when sighted individuals are blindfolded for only two days, their visual cortex activates when they perform tasks with their fingers or when they hear either tones or words. Two days is too short a time for new synapses to grow from touch and hearing areas into V1. More importantly, removing the blindfold for just 12 hours reverts V1 so that it responds again only to visual input. Thus, the brain's sudden ability to "see" with the fingers and ears may depend upon connections from other senses that are already there but are not used so long as the eyes input a signal. Such experiments show that we all harbor untapped multisensory potential.

Finally, sensory substitution[13] experiments underscore that latent cross-sensory connections exist in everyone. We normally think of the tongue as a taste organ, but it is loaded with touch receptors, making it an excellent brain–machine interface that can learn to "see" when a tingling electrode grid is laid on its surface.[14] The grid translates a video input into patterns of touch, allowing the tongue to discern qualities usually ascribed to vision, such as distance, shape, direction of movement, and size. The apparatus reminds us that we do not see with our eyes but rather with our brain. The technique was originally developed to assist the blind; a more recent application, however, feeds infrared or sonar input to the tongue grid so that divers can "see" in murky water and soldiers can have 360° vision in the dark. The implication of these applications is that synesthesia is a latent capacity in everyone.[15]

Examples like these all serve to illustrate that the normal brain is heavily cross-wired. The difference between the synesthetic and nonsynesthetic brain therefore is not *whether* there is cross talk, but rather *how much* there is.

Cross Talk in the Synesthetic Brain

The most plausible hypothesis for why synesthetic brains differ from non-synesthetic ones is this: synesthesia reflects an increased degree of cross talk between normally separated brain areas and the networks of which they are part.

But how do we know that synesthetes do not instead simply have a lowered threshold for being conscious of normally present cross talk? Is

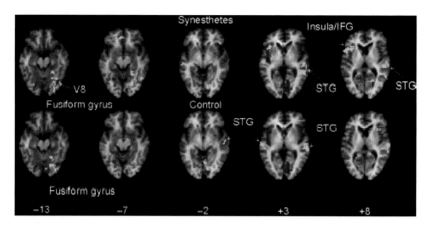

Figure 9.2
When a synesthete listens to words that trigger color experiences, the V4 color area
activates. (From Nunn et al. [2002], with permission.)

there a way to distinguish these hypotheses? If there actually is more cross
talk in synesthetes, we should be able to see evidence for it with neuroimaging. For example, when a word → color synesthete hears "three," which
triggers, say, magenta, can we measure increased activity in the visual color
areas of the brain? In 2002 Julia Nunn and colleagues set out to answer
this question.[16] They found that spoken words led to brain activation in
the region specialized for color vision, V4 (see figure 9.2).

Normally, the V4 complex is activated only when viewing real colors in
the outside world. The fact that other visual areas were not activated in
this study suggests that the activation of V4 alone is sufficient to consciously perceive color.[17]

However, a critic might ask, What if subjects were merely *thinking* about
a word–color association but not synesthetically experiencing any color?
Would that produce the same MRI results? To address this possibility,
several nonsynesthetes rehearsed word–color associations to learn them
by rote memorization. When these subjects were tested in the scanner,
they had no activation in the color region that was activated in synesthetes—so thinking about an association is not enough to explain the
results.

Finally, what if the synesthetes were merely better than controls at *imagining* colors? Can simply imagining magenta activate the color area? To

address this, the experimenters further instructed subjects to *imagine* colors. As expected from earlier studies,[18] doing so was insufficient to activate the color area.

Collectively, these findings suggest that from the brain's point of view the synesthetic experience of a color is less like imagination and more like actual color perception. Nunn's findings are important for several reasons. Besides confirming again the authenticity of the phenomenon for hard-nosed skeptics, it demonstrates that these kinds of synesthetes actually do have measurably increased cross talk in their brains.

Previous PET scanning in word → color synesthetes suggested that synesthetic color experience resulted from higher rather than lower visual areas—in other words, well into the thinking brain instead of its earliest sensory levels.[19] But the technically superior fMRI study by Nunn's team suggests the opposite, namely, that synesthetic experience is the result of low-level activation of brain areas specialized for basic properties (in this case, color).

Further studies have verified the conclusion that synesthetic experience activates the same brain areas that real experience does. Mere imagery, on the other hand, is insufficient to cause activation.[20] That is, if one is asked to merely *imagine* a color (e.g., "Is a canary darker yellow than a banana?"), a different set of areas become activated.

Following the success of the Nunn study, Ed Hubbard and colleagues turned their sights and their scanner on six grapheme → color synesthetes and six nonsynesthetic controls.[21] They hypothesized that when these synesthetes looked at a grapheme, the V4 color area would activate. To this end, they presented their subjects with white numbers and letters on gray backgrounds. They also included nonlinguistic symbols that did not elicit any color for the synesthetes. As predicted, they measured more activity in V4 in synesthetes than nonsynesthetic controls (see figure 9.3).

Hubbard and his colleagues further found a correlation between a synesthete's *behavior* on perceptual tasks and the size of the fMRI signal. Specifically, they found that the size of the fMRI signal in several visual areas roughly predicted how much the subject would be better than controls at texture segregation (say, picking out a triangle composed of 2s against a background of 5s), and whether they would have an advantage in identifying a "crowded" letter in the periphery (i.e., a letter surrounded by other flanking letters). In other words, the stronger the fMRI signal, the easier it

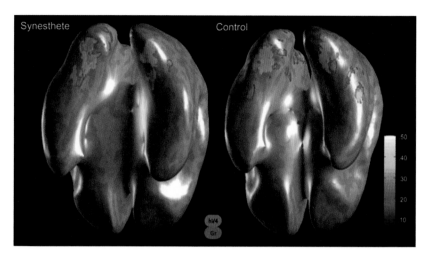

Figure 9.3
Functional magnetic resonance imaging data from six grapheme → color synesthetes. In this presentation, scanned brains are "inflated" by the computer so activity in the creases can be more easily seen. Human color area V4 is in purple, and the areas responsive to graphemes are shown in blue. Red indicates regions of increased activity when subjects view graphemes. Both synesthetes and controls show activation overlapping the grapheme area, but only synesthetes show additional activation in V4. (From Hubbard et al. [2005].)

was for the synesthete to segregate graphemes based on their synesthetic color.

Hubbard and colleagues were able to find these individual differences because, as we have emphasized throughout this book, synesthetes are a diverse group. But what do these differences mean? Does the size of the fMRI signal reflect the strength of the synesthetic experience, or rather does it mean that there was more than one *type* of synesthete in this study? This remains difficult to answer because of their small sample size (six subjects). Future experiments will have to tackle this question for a better understanding of the neural basis of the phenomenon. In the meantime, it is encouraging that not only can synesthesia be verified with neuroimaging but also the imaging results can be correlated with behavior.

The Involvement of More Areas

Despite the central prominence of area V4 in the studies above, we must emphasize that V4 is not the singular location of color synesthesia. It is easy to get excited by brain scans and mistakenly conclude that hot spots of activity explain the phenomenon. V4 accounts for only part of the overall synesthetic experience, namely, the color.

Certainly more than one structure is involved, if not least because we know that synesthesia is often a multifaceted experience. In the case of grapheme → color synesthesia, it seems at first glance sufficient to get only two areas talking to each other. However, even this simplest instance is deceiving because there is actually more to the experience than merely sensing color. After all, synesthetic experience includes an awareness of the percept, some level of affect toward it, and remembering it afterwards. Accordingly, components for attention, affect, and memory must be part of the circuit that underlies the experience.

The minimally necessary circuit is even more involved in the case of synesthetic forms such as [sound → color, location, movement]. Components required for this experience are (1) auditory cortex, (2) V4 for color, (3) parietal neurons for the spatial sense, (4) V5 for movement, (5) thalamus and cingulate cortex for attention, and (6) limbic components for affect. In reality, during a synesthetic experience that is part of an ongoing stream of consciousness, even more structures participate. Because synesthesia comes and goes, lasting for only a second or two, such networks are temporary collaborations among brain areas. The collaborations come into existence in response to a stimulus, then dynamically break apart as the experience fades. For reasons of technical limitations, our current technologies do not always allow us to spot these temporary networks in a scan.[22]

Leveraging Diversity to Understand the Brain

The fact that there are such large differences among synesthetes (some hear colors, others taste shapes, and so on) allows us to begin mapping out the different parts of the network involved in cross talk. By doing so, we can begin to see which areas are most likely to be talking to one another and which ones are never paired. This act of cartography allows us to put differences under the spotlight instead of sweeping individual differences under the rug, as a field does when it is young. We look at two types of individual differences in grapheme → color synesthesia to ferret out the brain areas involved in the synesthetic network.

Sensitivity to the Letter on the Page

In chapter 3 we introduced Hubbard and colleagues' finding of a synesthete for whom colors faded away when letters were presented at low contrast.[23] For this subject, a black letter on a white background triggered synesthetic color, as would a white letter on a black background—but a gray letter on a light gray background would not induce a color (or at least not with the same strength; see figure 9.4).

For this rare type of grapheme → color synesthete, the *concept* of a letter is not sufficient to trigger color. Instead, they are sensitive to the details of what is presented on the page. Hubbard and colleagues called such subjects "lower" synesthetes (indicating their low level of association), as opposed to the more common "higher" synesthetes whose color association depends on the concept more than the visual stimulus (see chapter 3).

This distinction between lower and higher synesthetes suggests a very direct hypothesis about the neural differences between the two groups. Let us first consider the lower synesthetes: in the brain's fusiform gyrus, an area dubbed the visual word form area (VWFA; see figure 9.5) responds to visually presented words, letters, and numbers.[24,25] This area does not encode for spoken words, nor to false fonts with equivalent visual

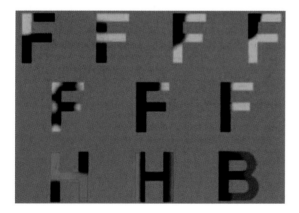

Figure 9.4

Example of a grapheme → color synesthete whose induced colors disappear when graphemes are presented at low contrast. (From Hubbard et al. [2006].) Letter F at 40%, 30%, 10%, 10% (on a second occasion), 5%, 4%, and 2% contrast levels; H at 30% and 5%; B at 30%.

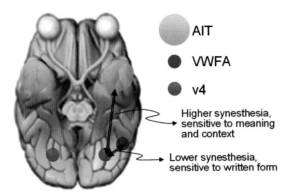

Figure 9.5
View of the brain from the underside with the cerebellum removed for a clear view of the visual area. Green = primary visual cortex. Arrows shows areas of increased cross talk. The visual word form area (VWFA) is mostly found only on the left hemisphere in humans and is thus pictured only on the left side here. AIT; anterior inferotemporal cortex.

complexity.[26] Intriguingly, this area responds less to low contrast than to high contrast letters.[27,28] Thus, the hypothesis suggested by Hubbard and colleagues is that in lower synesthetes, VWFA is directly cross talking to V4. By contrast, in higher synesthetes, a different brain area cross-activates with V4. Phil Merikle and colleagues in Toronto suggest that this brain area is the anterior inferior temporal (AIT) cortex, which encodes conceptual representations of words, letters, and numbers rather than the details of their visual forms.[29] The hypothesis that AIT cross talks with V4 is consistent with the fact that for the majority of synesthetes, colors are modulated by context and meaning.

In other words, some synesthesia is triggered by the percept (the letter on the page, implicating the VWFA), and some is triggered by the concept (the thought of the letter is sufficient, implicating area AIT).

This hypothesis makes certain predictions. First, it suggests that neither higher nor lower synesthetes will be sensitive to the capitalization of a letter (i.e., j and J will trigger the same color), because activity in VWFA and AIT is not sensitive to capitalization. Indeed, this insensitivity is widespread in synesthetes. On the other hand, an effect of font style might be expected in lower synesthetes, since VWFA might activate less to atypical fonts.[30] Two research groups have previously reported a small effect of font on synesthetic color,[31,32] but it remains to be seen whether they were sampling lower or higher synesthetes.

The hypothesis depicted in the figure introduces an important question. Since V4 and VWFA are so close to one another, why aren't most synesthetes of the lower rather than the higher variety? The fact that the lower variety is so rare indicates that proximity is not the key to understanding cross talk. Just because neural areas are neighbors does not mean they are more likely to have increased communication. Why distant areas (such as V4 and AIT) *are* likely to have more cross talk remains unknown.

What Does It Mean for Perceptions to be Localized?

We now turn the spotlight on another difference among grapheme → color synesthetes. In chapter 3 we introduced the difference between those synesthetes who experience color in a particular location ("localizers") and those whose color experience has no sense of location ("nonlocalizers"). These two types of color → grapheme synesthetes provide a good example of the way that different brain areas can interconnect. For localizers, at least a three-way interconnection is implicated among brain areas that code for (1) graphemes, (2) colors, and (3) spatial areas. These areas are respectively the (1) VWFA and/or AIT, (2) area V4, and (3) the parietal and hippocampal regions. For nonlocalizers, only a connection between the first two areas is implicated.

Moreover, [grapheme → color, location] synesthesia is not the only incidence involving synesthetic location. As we explored in chapter 5, a common form of synesthesia is spatializing number lines and other sequences. Regarding the numerical foundation of this kind of synesthesia, we highlighted the left angular gyrus as an important network node because the angular gyrus is important in tasks involving numbers and arithmetic. We turn now, however, to the issue of localization.

The neural representation of space is emerging as a large network of frontal, parietal, and hippocampal regions. Within the parietal lobe, the intraparietal sulcus (IPS) has received much attention in recent neuroimaging (see the green and yellow strip in figure 9.6). The IPS subdivides into several regions that involve different aspects of spatial representation. For example, the lateral intraparietal (LIP) area seems to represent the position of an object in an eye-centered frame of reference (see the red lines in figure 9.6). Nearby, the ventral intraparietal (VIP) area represents targets in

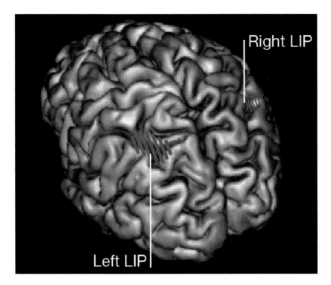

Figure 9.6
Approximate location of human lateral intraparietal (LIP) area is indicated by the red hashed region, with the posterior intraparietal sulcus (IPS) indicated in green. The anterior IPS is indicated in yellow. (From Hubbard et al. [2005].)

a head-centered frame of reference. VIP neurons have their receptive fields jointly determined by tactile and visual motion. Whereas area LIP codes for far space, area VIP codes for near space. Hand-centered coordinates, crucial for fine grasping, are represented in anterior intraparietal cortex (see the yellow line in figure 9.6). Finally, three-dimensional shape is calculated by the caudal intraparietal area.[33]

The point of this laundry list of areas and their types of spatial encodings is to illustrate that spatial encoding is not one thing in the brain. Instead, there are many aspects to space—similarly, there are many varieties of spatial synesthesias. Note that upon close examination, grapheme–color localizers do not all give the same description of how they localize. Some report a synesthetic color "on top of the letter, like a transparent overlay," while others see the color in a location separate from the letter. Future studies will tease out these differences in terms of the neural networks involved. Similarly with number forms, one finds very different descriptions of localizations. Some synesthetes describe being able to "move around" their number forms, whereas for others the form is fixed in body-centered coordinates. Some can zoom in and out of their forms while others stand at a fixed distance in relation to it. For some the form is distant, and for others near. We expect future synesthesia studies to refine our definitions of these different types of synesthesia based on the details of the brain areas involved. As we emphasize, most studies of synesthesia until now have ignored individual differences, but the time has arrived to address them.

The Development of Cross-Activation

We have seen that neuroimaging demonstrates increased cross talk in synesthesia. The next question is how cross talk comes about. There are essentially two theories: more connections or less inhibition. We also introduce a third idea: reduced plasticity during a critical period. We address each in turn, below.

More Connections

The fetus makes two million synapses each second, leaving newborns with an excess of working connections between brain areas. The overwiring hypothesis of synesthesia stipulates that these excess connections are insufficiently pruned. That is, a synesthetic brain is like a garden whose shoots are not properly trimmed back.[34,35,36] The proposal is that the persistence in adults of these extra connections leads to synesthesia.

A variant of this idea is that there is increased *outgrowth* of neurons in a synesthete's brain, like a garden that persistently sends shoots everywhere, yet despite regular pruning remains massively overgrown. What both these ideas (insufficient pruning and increased arborization) have in common is the idea that a synesthetic brain simply harbors more synaptic connections than a normal brain does. (Incidentally, McMaster University neonatologist Daphne Maurer suggests that if increased wiring is the correct explanation for synesthesia, then perhaps all neonates are synesthetic, only to lose it around three months of age.)[37]

If the increased wiring hypothesis is correct, we should expect to see synesthesia present from birth onwards, which we don't. Instead, synesthesia does not become apparent until mid-childhood. For example, grapheme-based synesthesia does not appear until after age three, and emotionally mediated synesthesia between the ages of three and five.[38]

At present, we have not encountered enough childhood cases to reliably know the earliest age at which different kinds of synesthesia manifest. We will say more about the timing of synesthesia after discussing its genetics and the interaction of nature with nurture.

Less Inhibition

Although excess wiring has received the most attention, a second view, first suggested by Peter Grossenbacher,[39] implicates faulty *inhibition* between brain areas as the cause of synesthesia. The idea is that in normal brains, excitation is balanced by inhibition, whereas in synesthetic brains, the inhibition is diminished. In this framework, the same amount of cross-sensory wiring exists in everyone, but it is typically unable to exert a strong effect because it is counteracted by inhibition[40] (see figure 9.7). The disinhibited structure may be nearby or remote. What matters is that connections exist between the two entities.

One finding favoring the disinhibition idea is that forms of synesthesia occur in nonsynesthetes under certain conditions—we call these "acquired" synesthesias. For example, synesthetic experiences are not uncommon during meditation, in states of deep absorption, or as otherwise nonsynesthetic individuals fall asleep. One nonsynesthete, Pat C, reports that if a door slams just as she is drifting off, she sees a burst of colors. Normally

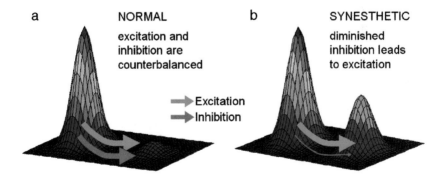

Figure 9.7
In this cartoon of neural connectivity, green represents excitatory connections and red represents inhibitory connections. (*a*) When inhibition functions normally, a bulge of activity in one area does not spread because the excitation and inhibition are counterbalanced. (*b*) When inhibition is diminished, activity in one area can lead to activity in another.

she does not experience sound → color synesthesia. Something about the state of her brain as it enters sleep allows her existing connections to change their functional relationships. Another example is the fact that LSD and similar hallucinogenic drugs can induce synesthetic experiences in nonsynesthetes.[41] Still another concerns newly blind individuals learning Braille and sighted individuals being blindfolded; both reassign their visual cortex to touch and hearing. These examples indicate that cross-wiring is present in everyone but that activity in one region does not normally kindle activity in another region because excitation and inhibition are balanced.

If this theory is true, claim Peter Grossenbacher and Chris Lovelace, then "synesthesia is entirely mediated by neural connections that exist in normal adult human brains."[42] The connections are there *anatomically* in everyone, but they do not necessarily *function*.

Less Plasticity

Let's imagine that a red J is seen by a young child—perhaps a refrigerator magnet, a sign on an elementary school wall, or a self-made choice using a box of crayons. A basic rule by which neurons modify the strength of their connections is this: those neurons that are active at the same time (say, those coding for J and red) will strengthen their connection. This rule of "fire together, wire together" is known as a "Hebbian rule" after Dr. Donald Hebb. For most people, the connection between J and color will continue to be modified with each new sighting of the letter J in different colors. Thus, when a green J is seen, the connection between J and green is strengthened and the connection between J and red is weakened. With enough exposure to differently colored Js, the letter–color pairings will average out, leaving no particular association.

Given this background, one possibility is that synesthetes have slightly reduced plasticity.[43] "Plasticity" refers to the ability to modify a connection once it has been set. In other words, once an initial letter–color pairing has been set, it sticks. Note that a consequence of this theory is that excessive exposure to individual letter–color pairings should be able to induce synesthesia in any child's brain. In weighing the case for these different theories of synesthesia—more connections, less inhibition, and reduced plasticity—we next turn to the strange cases of synesthesias acquired later in life.

Acquired Synesthesia: Same or Different?

A number of acquired synesthesias involve sound → sight coupling. Inducers include LSD, hallucinations resulting from sensory deprivation, meditative states, closed head trauma, and temporal lobe epilepsy (TLE). We examine each in greater detail.

LSD-Induced Synesthesia

LSD, mescal, and other anti-serotonergic hallucinogens sometimes produce synesthesia, particularly the sound-to-sight variety. Almost all measures of color vision are affected in volunteers given LSD. Given the drug's rapid onset of action, we conclude that it must operate on preexisting pathways. LSD is thought to have two main effects: increasing transmission of primary sensory inputs at early levels while inhibiting pathways within the cortex.[44]

In general, serotonin dampens overactive neurons to various internal and external stimuli. Given that serotonin is primarily an inhibitory neurotransmitter and LSD blocks its receptors, the resulting disinhibition allows neural targets to be more easily activated by abnormal inputs, hence synesthesia. Serotonin receptors are maximally concentrated in the hippocampus, thalamic nuclei, basal ganglia, and cerebral cortex.[45] Depth electrodes implanted in animals and humans while on LSD reveal a desynchronized cortex and synchronized, paroxysmal discharges in hippocampus and amygdala. This means that such individuals are aroused but unable to discriminate among sensory channels. In humans, these subcortical discharges coincide behaviorally with heightened emotion and perceptual distortion during the drug high.[46]

How similar is the experience of LSD-induced synesthesia to the naturally occurring kind? Volunteers given LSD are in some ways reminiscent of synesthetes in their tendency toward concreteness, the vividness of the percepts, the heightened memory of them, and the emotional freighting of the experience. However, the experiences are said to differ according to five synesthetes who have tried LSD in the past. Four experienced no synesthesia, and one (MM) felt his spontaneous synesthesia was intensified. Additionally, Michael Watson experienced a very unpleasant "sensory overload" while on LSD, but no synesthesia.

It is possible that since these subjects naturally experience synesthesia, LSD did not add to it. A study that could address the relationship between natural and drug-induced varieties would be the following: nonsynesthetic volunteers on LSD could associate heard sounds with colors both while on LSD and off LSD. The subjects would take the same type of consistency tests described in chapter 3, and one could analyze the results to answer whether their sound–color pairings are internally consistent while on the drug but less consistent when off the drug. Then, six months later, the same subjects could be retested on the drug. If their sound–color pairings match those chosen the first time, it would be strong evidence that a subconscious landscape of associations, present in everyone, is able to be unmasked by the disinhibition caused by LSD. In contrast, a negative result on this test would suggest that naturally occurring and drug-induced synesthesia are fundamentally different.

Sensory Deprivation and Release Hallucinations

A sensory-deprived brain starts projecting a reality of its own, perceiving things that are not there. The situation is not so rare as it might first seem. A common experience occurs when your hearing is deprived by the white noise of the shower: in that situation how often have you hallucinated that the phone was ringing or that someone was calling your name?

A similar situation happens with the loss of hearing, touch, or vision. Even simple boredom can produce hallucinations.[47] With increasing loss of input, sensory-deprivation hallucinations progress in degree of severity. At first, visual hallucinations are simple and consistent (geometric patterns, mosaics, lines, rows of dots; see Klüver's form constants in chapter 8), progressively becoming more complex and dreamlike as they involve bizarre juxtapositions of people and objects.

Hallucinations that are "seen," "heard," or "felt" in a blind, deaf, or insensate field are called release "hallucinations," as if a given sensory cortex were uncoupled and released from its normal upstream input to become active on its own.[48] For example, one individual with brain damage that caused blindness in his left visual field had three kinds of release hallucinations: *generic synesthesias* of perpendicular red-and-green lines, red-and-blue spots, and black-and-white pulsations; *metamorphopsia*, in which the right half of faces seems to melt and take on a yellow or violet tint; and *palinopsia*, in which visual images appear to multiply, repeat, or leave trails.[49]

As mentioned above, Armel and Ramachandran reported an individual with retinitis pigmentosa who steadily lost vision since childhood until becoming completely blind at age 40.[50] A few years later he started experiencing touch → vision synesthesia. The touch threshold for having synesthesia was higher than that for simply perceiving touch. Similarly, individuals with progressive hearing loss can experience musical and verbal hallucinations.[51]

Patients with lesions in the optic nerve or tract can see sound-induced photisms in their blind fields that often startle them.[52] The sounds are usually those of daily life and include the clanking of a radiator, the crackling of a wall as it cools at night, the whoosh of a furnace ignition, a dog's bark, and slamming doors (see table 9.1). Photisms range from simple flashes of white light to colored forms that look like a flame, an amoeba, oscillating flower petals, a spray of bright dots, or kaleidoscopic effects. All last only "a split second," and some patients experience multiple photisms whereas other have only one.

Table 9.1
Characteristics of sound-induced photisms

No.	Photism appearance	Color	Location	Sounds
1	Flame, flashbulb	Red–orange, white	In scotoma	Not sure (sharp)
2	Spray, pollywogs, kaleidoscope	White, pink, red, black, green	In and out of scotoma	Clap, CT scanner
3	Flash	White–yellow	In scotoma	Walls crackling, digital clock, TV crackling
4	Lightbulb	White–blue	In scotoma	Not sure (soft)
5	Flash	White	In and out of scotoma	Engines, loud sounds
6	Flashbulb	White	In scotoma	Electric blanket, digital clock
7	Petal, amoeba, goldfish	Pink, white, yellow	In and out of scotoma	Furnace, dog, voices, clatter
8	Plaid	Green	In scotoma	Book or fist on desk, loud
9	Flashbulb	Pink	In and out of scotoma	Furnace, door slam, TV, radio

(From Jacobs et al. [1981], with permission.)

Note. A scotoma is a blind spot in the visual field.

It is peculiarly interesting that the photisms are perceived to arise in one eye and be induced by sounds that are heard only with the ear on that side. This is inconsistent with our usual understanding of vision and hearing anatomy. We are unable to tell which eye sees an object unless we cover first one and then the other to determine that only one eye, in fact, sees a certain object (such as when the nose is in the way or when objects appear in the nonoverlapping fields at the far edges). Similarly, acoustic localization of objects depends on differences in the sound reaching both ears. Yet, in a case described by Jacobs and colleagues,

... the click of an electric blanket thermostat induced a flashbulb photism in the right eye of patient 6 only when the thermostat located to her right clicked; the same clicking from her husband's thermostat located to the left never induced the phenomenon. A petal photism was perceived coming from the right eye of patient 7 when a nurse spoke into his right ear. The photism never occurred when the nurse spoke into his left ear.[53]

Such spontaneous visual experiences are common, occurring in 60% of individuals with damage to the visual pathways.[54] Physicians typically need to specifically ask about such experiences, because patients are reluctant to spontaneously reveal "crazy" symptoms.

Individuals with damage to the nerves connecting their eyes to the rest of their visual system often end up with supersensitivity of downstream structures, including areas that bring together several senses ("multisensory" areas). And such sensitivity to these areas, or even direct damage to these areas, can lead to phenomena that resemble synesthesia. In one case, a brainstem tumor encroaching on the left medial temporal lobe caused acquired [sound → color, shape, location, movement] synesthesia that stopped once the tumor was removed.[55] The synesthesia was stimulus dependent and could be manipulated: increasing the click rate at a constant volume altered the intensity and movement illusion of the individual's photisms.

Meditative States

Formal meditative states such as occur in the practice of Zen or Yoga are states of reduced input. Thus, they are qualitatively akin to states of sensory deprivation. In asking whether synesthesia can be cultivated, psychiatrist Roger Walsh at the University of California turned to three groups of

Buddhist meditators with different lengths of practice: Tibetan retreat participants, physicians in a Vipassana meditation group, and adept teachers from three Buddhist schools (Theravadin, Tibetan, and Zen).

Walsh claims that synesthesia was experienced by 35%, 63%, and 86% of meditators in each respective group. Length of meditative practice correlated with an increase in synesthetic experience. As a group, for example, retreat participants were most naive; yet within that group those who did experience synesthesia had nearly twice as much practice time as nonsynesthetes, a difference significant for years of practice rather than time spent in retreat. Compared to the baseline incidence of 4%, synesthesia is about ten or more times as common during meditation. So don't take LSD, but learn to meditate if you hope to experience it. Incidentally, within the most experienced teachers group, 57% of synesthetic experiences were multisensory.

Walsh points out that meditation has been experimentally shown to enhance perceptual sensitivity.[56] In *The Man Who Tasted Shapes*, Richard argued that "synesthesia is actually a normal brain function in every one of us, but its workings reach consciousness in only a handful."[57] Walsh contends, based on his empirical observations, that awareness-enhancing techniques such as meditation may unmask an ever-present synesthesia to consciousness.

Walsh's most intriguing observation is that the most experienced meditators report concept-based or categorical–sensory amalgamations. That is, cognition such as "emotions, thoughts, and images" are experienced in sensual terms such as sound, taste, or touch. For example, emotions are most often experienced as synesthetic touch, less so as taste or sound. One participant "tasted thoughts" whereas another felt them as quivering "vibrations," while for a third, "the thought of a friend can have the scent of frangipani." Whether these reports qualify as the type of synesthesias we discuss in this book remains an open question.

In the liturgy of Soto Zen, the sandokai scripture states, "Each sense gate and its object all together enter thus into mutual relations," whereas the scripture of great wisdom (prajnaparamita) asserts there is no distinction among the senses and concepts:

O Shariputra, form is only pure,
Pure is all form; there is, then, nothing more than this,

For what is form is pure—and what is pure is form;
The same is also true of all sensation—thought, activity, and consciousness . . .
O Shariputra—in this pure there is no form, sensation, thought, activity, or
consciousness;
No eye, ear, nose, tongue, body, mind; no form, no tastes, sound, color, touch, or
objects . . .[58]

Walsh comments, "To what extent these ancient claims represent accurate descriptions of very advanced meditation experiences, and to what extent they represent idealized extrapolations is unknown." His observations, however, suggest that meditators may be an untapped pool for studying synesthesia and cross-modal metaphors.

Temporal Lobe Epilepsy

Seizures originating in the temporal lobe are often called complex partial or psychomotor epilepsy. (First, a clarification of terminology: a "convulsion" is a violent muscular contraction whereas "seizure" refers to a sudden electrical discharge in the brain. Not all seizures cause muscular convulsions.)

Unlike grand mal seizures that involve an electrical storm throughout the entire brain (and therefore cause widespread motor convulsions), the seizure discharge in TLE is circumscribed. Because all association areas project to the temporal lobe, sensations and perceptions that are laden with affect can be the sole manifestation of this kind of seizure. Quite often afflicted individuals do not have any convulsions or abnormal motor activity at all; the seizure may be experienced entirely as an alteration in perception, thought, or feeling. Hence another name for TLE is "psychic seizure."

One individual is described as having a threefold epileptic synesthesia involving vision, hearing, and pain. He heard the word "five" in both ears and saw the numeral 5 projected on a gray background while feeling a shooting pain in his face.[59] His EEG showed left temporal spikes. A similar individual experienced "visual pain and visual hearing."[60] Although not epileptic in nature, synesthesia also follows closed head trauma. Richard reported that 1.4% of individuals with concussion experienced synesthetic pain, lasting several months, in the head, neck, and arm in response to bright light or loud noise.[61]

Tastes and smells occur in about 4% of temporal lobe seizures. Tastes are usually not described in detail but in general terms, such as "bitter," "unpleasant," or simply "a taste," unless seizures extend beyond the temporal lobe. In that case the taste becomes more specific ("rusty iron," "oysters," "artichoke") the way that tasted phonemes do. Seizures originating in the temporal lobe are intermixed with many subjective symptoms, a finding that implicates the anterior temporal cortex and underlying limbic structures in these synesthetic experiences.

The following epileptic synesthesias are from a series of cases and show how a seizure can cross-activate different functions:[62]

Case 21 A taste of bile, tingling of the left wrist, twitching of the left corner of the mouth, and jerking contractions of the left side of the body.

Case 24 Stomach pain, shivers, a bitter taste, nausea.

Case 25 A lump in the throat, oral movements, photisms in the right upper field, a bitter taste.

Case 28 An intense heat that ascends from the stomach to the mouth accompanied by a disagreeable taste.

Case 30 Bitter taste, extreme salivation, swallowing, spitting (sometimes vomiting), angry outbursts accompanied by shouting.

Unrelated to the above collection of cases, a twenty-three-year-old man had a "rough, bitter sensation in his mouth" combined with a "peculiar sensation passing down the right arm into the hand, then up the spine from the level of the shoulders to the head, finally spreading over the back of the skull as a cold sensation." He would often shiver during the seizure and repeatedly see a scene from his childhood. He was found to have a large pituitary tumor that undermined the posterior three-fourths of the temporal lobe.

Synesthete Richard N has up to twenty temporal lobe seizures a day, a number which is not uncommon. While he says that "everything has its own color, texture, and sometimes smell," he illustrates how both generic synesthesias and more elaborate epileptic hallucinations combine:

It's the colors and the music that are the meat of the episode. But I also see people, hear voices, and see places all at the same time with the music. Aside from the beauty of the light and sound show my brain is putting on, the physical sensation is ecstasy. It's the only word that describes it.

After it is over, I begin to sweat profusely all over and my heart races as if I had just done strenuous exercise. . . . And I can never wait for the next one.

It remains to be determined whether epileptic synesthesias display the same consistency and are caused by the same neural mechanisms as naturally occurring synesthesia.

Chemicals and Synesthesia

If synesthesia is brought about by disinhibition, we expect particular substances that affect brain chemistry to modulate synesthesia in specific ways. And indeed they do, as we discussed in chapter 6 with respect to Michael Watson, who was given alcohol, amphetamine, and amyl nitrite (refer back to table 6.3). Individuals who report an effect often find their synesthesia enhanced by alcohol, fatigue, high emotions, or marijuana and reduced by caffeine, cigarettes, or antidepressants. While these correlations do not hold uniformly for all synesthetes, they are generally consistent with the hypothesis that decreased inhibition does lead to increased synesthetic experience.

The Genetics of Synesthesia

In a book written in 1883, Sir Francis Galton pointed out that synesthesia runs in families[63] and suggested that the condition is heritable. A century later in 1989, in the first edition of *Synesthesia: A Union of The Senses*, Richard Cytowic examined the inheritance patterns of eight families and proposed that synesthesia is transmitted as a dominant trait. The salient feature supporting dominance was the trait's appearance in multiple individuals within a generation as well as between successive generations.

In 2005, Ward and Simner looked at the inheritance patterns in seventy-two families (of varying size) and conjectured that the trait may be inherited on the X chromosome.[64] This is because the trait seemed to pass along in a pattern characteristic of X-linked traits: it is passed from mothers (XX) to either sons (XY) or daughters (XX) or from fathers to daughters—but not from fathers to sons.

Time has turned up instances of father-to-son transmission, which are problematic for the X-linked story because a single authentic case invalidates it. However, such findings by themselves are not enough to rule out X linkage, for two reasons. First, it may be that the trait actually passes down from the mother who is a silent carrier of the gene but does not express it herself. Second, infrequent spontaneous mutations lead to the synesthesia gene. (Incidentally, either of these explanations can also address the instance of a synesthete for whom neither parent appears to be synesthetic. In this case the trait can be passed along silently or it can arise spontaneously as a new mutation.)

More damning to the X-linkage hypothesis is the fact that preliminary genetic data do not support it. As of this writing, David Eagleman's laboratory has collected DNA from over 100 members of several different synesthetic pedigrees (see figure 9.8). Family linkage analysis, a form of genetic analysis that identifies a region in the genome that carries important changes, suggests a region critical for synesthesia on chromosome 16, but not on the X chromosome.[65]

Whichever chromosome(s) house the gene(s) for synesthesia, a phenomenon called "incomplete penetrance" appears to characterize the transmission. "Penetrance" is the relative ability of a gene to produce an effect in the organism. In synesthesia the effect is illustrated by identical twins

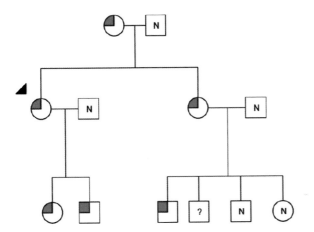

Figure 9.8
A representative family tree showing inheritance of synesthesia across generations, and in multiple individuals within a generation. A red quadrant marks synesthetes, N stands for "nonaffected," and a question mark denotes those who have not yet been tested. Data are from David Eagleman's laboratory.

wherein one twin is synesthetic while the other is not.[66] Since we already have evidence that the trait is passed genetically, such a finding highlights the fact that not every carrier of the gene expresses the trait. Such a thing happens in many genetic conditions. For example, if you come from a family of left-handers, you are more likely to be left-handed than not, but not guaranteed to be. Likewise, if you have a left-handed identical twin, it does not necessarily mean you must be left-handed.[67] Carrying a gene does not always mean it will be expressed or predictably influence an associated physical trait. This is what is meant by "incomplete penetrance."

We now know thanks to the random population studies of Julia Simner and her colleagues[68] that male and female synesthetes actually occur in approximately equal numbers. Thus, historical debate regarding whether a synesthesia gene could be lethal to male embryos[69] has been rendered obsolete.

Who Else Might Harbor the Gene?

So far, we see that if the localized expression of the synesthesia gene happens to lead to increased communication between hearing and color areas, for example, then a person will have sound → color synesthesia. However, we know from foregoing chapters that the synesthesia gene can be expressed simultaneously in three, four, or five locations. If the gene were expressed only in nonsensory areas, perhaps leading to increased communication between regions of the frontal lobe, a person would not be synesthetic by our current definitions and perceptual tests. In such as person different *sensory* traits such as hearing and vision would not be cross-activating. Instead, areas involved in reasoning, planning, decision making, and so on would cross-activate. What would such a phenotype look like? A genius, an artist, or a madman? We do not yet know, although we suggested in chapter 8, following Ramachandran and Hubbard, that the gene is an attractive candidate for the neurological basis of metaphor and the ability of highly creative people to see connections between seemingly unrelated things.

Therefore, when we think about penetrance (what fraction of those carrying the synesthesia gene express the trait), we find at least two ways to carry the gene silently. One can fail to express the gene (perhaps by inactivating the relevant one of the two chromosomes) or express it in a brain region that makes it undetectable to synesthesia researchers who hunt only for sensory manifestations. If we can successfully find the gene that corre-lates with synesthesia, our next task will be to find out who else carries the gene in the general population and determine what unexpected cogni-tive or emotional traits they may possess.

Throughout this book, we use the term "synesthesia gene." But we have done so only in the interest of brevity. Our shorthand encompasses the possibility that synesthesia may well result from several genes, as well as the likelihood that different genes may be implicated in different families. These genetic details are the subject of current study in David's laboratory.

Why Are There So Many Forms of Synesthesia?

However many genes are found to correspond to synesthesia, one possibil-ity is that it is the *location* of their expression that gives rise to the many different kinds of synesthesia. (This may also explain why several different forms are often present within a given family.)

Vladimir Nabokov, for example, experienced colors in response to the sounds of language, whereas his mother experienced colors in response to music. In a more florid example, each of Amy P's synesthetic relatives has different kinds of synesthesia:

My dad has number and year forms, as well as coloured letters and days of the week. One sister has just a few forms (week and year) and . . . coloured letters. I have no colours, but have more forms than any of them (numbers, days, weeks, years, ages, centuries, alphabet), and my numbers have personalities and gender. The sister who used to have it [had] coloured letters, number personalities . . . and coloured smells.

When only color and letter areas of the brain are affected, a synesthete experiences grapheme → color. When spatial areas are additionally involved, synesthetic expression becomes [grapheme → color, location]. With increasing functional connectivity, possible synesthetic forms extend in richness—for example, [grapheme → color, location, personification, gender] and so on. Given the richness of brain specialization and the enormous number of possible combinations in which networks can inter-connect, it is clear why there is enormous variety in the expression of synesthesia. The 2004 meeting of the American Synesthesia Association reported 152 recorded forms of synesthesia. If one cares to continue cataloging varieties, the number will certainly climb higher. It is all a matter of where and how diffusely in the brain the gene exerts its effect.

Nature and Nurture

Synesthesia is shaped by nurture as well as nature, as we indicated in chapter 2 when discussing the progression of skills that babies naturally acquire. The environment does not influence all genetic traits, however. Eye or hair color, for instance, is going to fixedly express itself no matter how your parents treat you or what culture they expose you to. However, many genetically based physical traits do interact with the environment. Take, for example, height and weight. Although you arrive in this world with a genetic predisposition toward a given stature and heaviness, the environment is a major player in how tall and trim you actually turn out to be. The same interaction may well be true for synesthesia.

As shown in table 2.2, there is an orderly progression in which infants learn phonemes, word fragments, the names of what they eat, primary and secondary colors, how to tell time, how to distinguish one day of the week from another, and so on. The infant brain constantly changes and reorganizes itself as it experiences the world. This is called learning, which manifests as a physical change. Both genetically programmed brain reorganization and that due to unique experience continues robustly until about age eleven, has another spurt during puberty, and then tapers off around age twenty to twenty-five. At that point the brain mainly tinkers with insulation and the fine arborization of synapses in executive and integrative areas. (Lest we leave the wrong impression, brain plasticity does continue throughout adulthood, but at a much slower pace.)[70]

The fixed order in which an infant acquires skills suggests a natural research question, namely, is there a time window during which gene(s) involved in synesthesia exert their effect? We can answer this question by seeing if, in individuals with multiple synesthesias, certain kinds cluster together more often than by chance. That is, is someone who tastes phonemes also likely to taste colors, or is someone who personifies graphemes also likely to have emotionally mediated synesthesia? If such clusters occur, it suggests there is a critical period in a given synesthete during which the gene expresses itself by reducing feedback inhibition among affected brain areas. At present, synesthesia does not appear to be switched on at birth but rather becomes evident as children *acquire* skills such as knowing the names of colors, letters, and numerals or learn to *label* piano keys, time units such as days of the week, and so on. This is the learning that any theory of synesthesia must take into account.

It is fascinating that synesthesia is biologically based in genes and brains but almost always requires early life exposure to culturally learned artifacts such as graphemes, colors, food categories, and time units. This is arguably true even in instances of colored hearing, including those instances in which superficially there appears nothing much to learn, such as [environmental sounds → color, shape, etc.]. We say this because at some point in childhood an individual who has never attended to the sound of rain, for example, *learns* once and ever after what rain sounds like. That is, he or she *labels* it, and the label has meaning. In neuropsychology we know that labels need not be verbal (e.g., mute individuals and preverbal children demonstrate that they understand the meaning of various sounds). Although the question is arguably open in synesthetes, perhaps when they cognitively label an event, it either undoes nascent inhibition or prevents it from taking hold, hence synesthesia.

We mentioned that an extreme minority of synesthetes shows evidence of imprinting. That is, early life exposure to colors and patterns directly shapes their synesthetic experience, which then becomes locked in and stable over time. In one remarkable case in which the subject's mother saved his alphabet-shaped refrigerator magnets, their colors directly correspond to the individual's grapheme colors.[71] In other cases, the colored label on a child's piano keys determined her synesthetic color association with numbers and pitch, and in Joseph Long's case (discussed in chapter 4), the physical key labeled "middle C" was blue regardless of its sounding pitch, which depended on how accurately a given piano was tuned.

What is maddening to brain researchers is that no *single* explanation stands out as most plausible. And certainly none explain *every* kind of synesthesia. Yet many approach synesthesia thinking, "How hard could this be to understand? It's just crossed wires, isn't it?" As we see, the answer is much more nuanced.

It is often the case in science that examining a phenomenon raises more questions—and more fundamental ones—than it answers. This certainly has been our experience in studying synesthesia over our careers. In the past decade a team of scientists has explained an enormous amount of synesthesia's fascinating behavior while at the same time exposing very fundamental questions about how the human brain is organized.

Open Questions

We began with an orthodox view of brain organization in which specialized modules in the synesthetic brain are more cross-activated than they are in normal brains. It is clear from ongoing research as well as orthodoxy's failure to explain synesthesia fully that this picture requires refinement. The bottom line is that we doubt the brain is truly made up of modules. That is, the received wisdom that given discrete functions operate in separate channels is not holding up, and synesthesia is forcing a paradigm shift.

In the orthodox view of neuroscience, the brain is genetically prespecified as a collection of modules that embody the available tricks which evolution has discovered for detecting information about the world. In 1908, the German anatomist Korbinian Brodmann made the first attempt to decompose the cortex into modules based on microscopic distinctions in the way cells were arranged in a given area. The technical term for this approach is "cytoarchitecture." On this basis early ideas of brain function were inspired by the industrial metaphor, in which brain areas took in input, processed it, and spit out the result for the next station on the assembly line.[72]

By the middle of the last century, neurophysiology had taken up this theme, delineating distinct areas related to Brodmann's maps that seemed important in the processing of faces, color, motion, motor planning, executive action, intention, and so on. By the 1990s, it became clear that distinct brain areas do activate during fMRI studies when people perform particular tasks. Often, brain damage to the same areas leads to deficits in performing those tasks. Yet we stress that neuroimaging can sometimes be misleading because it emphasizes areas of peak activation and does not capture all the entities participating in the network that underlies a given function. In other words, scanning shows us a partial functional landscape, at best, and says nothing about the active structures that are participating beneath the threshold.

There is also another way of transmitting information throughout the brain as well as the entire body. It is called "volume transmission" and relies on the transfer of information by small molecules such as hormones, peptides, and gases. Volume transmission operates over a wide range of distances and speeds. If you think of the traditional hardwiring of neurons and synapses as a train traveling down a track, then volume transmission is the train leaving the track. The orthodox view rarely, if ever, takes volume transmission into account.

Yet to many neuroscientists, the idea of a modular brain seems natural— that is, modules are genetically specified, isolated from other modules (encapsulated in scientific lingo), and selected for by evolution. However, the modular arrangement breaks down for many reasons. First, it is clear that the brain's different so-called modules send and receive a great deal of information among them—which is simply not allowed by modules that, by definition, are encapsulated. For example, auditory cortex receives visual information, and the cortical area specialized for motion also responds to color. Even the primary visual cortex is heavily influenced by the organism's goals.[73] That is, what a so-called module does varies with context and the nature of the task occurring at a given time. For example, many neurons in parietal cortex seem to specialize in visual objects if the task calls for it—but as soon as the task calls for auditory expertise, these same neurons act like auditory neurons.

This does not mean neurons in the primary visual cortex are not visual but rather that they are not *purely* visual. Not only are these areas not isolated but they also are not dedicated to a fixed function. Instead, they adapt to the task at hand.[74] The so-called motion area (called MT, for middle temporal) appears specialized to detect motion, but its cells also respond to color input when color is needed to solve a problem. There are even cells in primary visual cortex that are activated by eye-movement preparation, while cells in the lateral geniculate, long believed to be a purely visual structure, turn out to be modulated by attention, body sensations, and task specificity. We do not yet know how cortical neurons adapt and change as a function of the required behavior.

We wish to emphasize that activation of an area on fMRI is certainly not evidence for a module, given that brain areas light up in response to stimuli that have no evolutionary basis for a module. For example, fMRI studies disclose regions distinctly activated by exposure to pronounceable nonsense words or even headless bodies. It is hard to make an evolutionary argument for why such modules would exist, so it encourages us to abandon the idea of assigning every spot of fMRI activity to a genetically specified modular function.

We believe the right way to think about brains is in terms of developmental programs working in concert with rich worldly interactions, which dynamically leads to neural specialization. Division of labor is something that develops from local border wars on the cortical map more than by genetically determining the specificity of modules. How this modifies our conception of synesthesia is something we explore in the next chapter.

Conclusions

Synesthesia is not localized in any one spot in the brain. Rather, we need to think in terms of neural networks that connect several brain structures, all of which contribute to the synesthetic experience. Thus, instead of being localized in the sense of classical neurology, we say that synesthesia exists as the dominant process in its distributed network at a given time. The normal human brain is heavily subspecialized, and a constant cross talk among specialized areas allows them to collaborate. The difference between synesthetic and nonsynesthetic brains is not whether cross talk exists but rather its degree.

Why does the synesthetic brain have more cross talk? It remains to be determined whether synesthesia results from increased connectivity (from a lack of pruning, for example) or simply from an imbalance of inhibition that leads to more cross talk within pathways that normally exist in everyone. It is still unclear why synesthesia is more common in children and sometimes fades away as the brain matures. Worst of all, the role of learning has not been incorporated into any theory. Hence, they all remain incomplete.

Depending on both genetics and learning over a lifetime, there are a vast number of possible combinations in which brain areas may interconnect. Depending on the fine details of interconnections, a person may express different forms of synesthesia. Sometimes these differences will be subtle, as in grapheme → color versus [grapheme → color, location]. In such cases the distinctions must be teased out with careful questioning. In other cases differences are obvious, as in smell → color versus number →

personality. Most studies of synesthesia until now have ignored individual differences, but we suggest leveraging the details of differences to discriminate exactly which brain areas are likely to be involved in which synesthetes. Whereas the variety of synesthetic expression is as varied as personality types, it is possible that a single gene underlies all these synesthetic forms that on the surface appear so varied. This is because synesthesia, in its various guises, follows particular patterns of inheritance as it runs through family trees. As we write this, the hunt for the synesthesia gene is in full swing.

10 Questions Ahead

In the past decade, synesthesia has moved from a curiosity greeted with skepticism to a subject of intense study in neuroscience, psychology, and genetics. This change comes from the confluence of two factors. First, the introduction of new technologies has allowed synesthesia to be rigorously verified for the first time. These technologies include fMRI, computerized behavioral tests (measuring consistency, reaction times, and memory), and a revolution in human genetics. The pace of study has also been hastened by the Internet, which has sped the dissemination of information and introduced tools such as Sean Day's e-mail lists and the online test batteries at www.synesthete.org.[1] At the time of this writing, synesthete.org has launched several translations, extending the reach of research to other languages and alphabets.

The new technologies have worked hand in hand with a second factor that has led to synesthesia's rise in science: a renewed interest in private, subjective experience, also known as consciousness. Until recently, consciousness was forbidden territory in neuroscience. The spirit of the field had been dominated for decades by the behaviorist school of thinking, which was headed by the American psychologist B. F. Skinner. Behaviorism asserted that consciousness was an unimportant illusion in a stimulus–response machine. It took scientists with the gravitas of Francis Crick and Christof Koch, among others, to establish consciousness as a real scientific problem.[2] After all, it *feels* like something to have pain. It *feels* like something to see the color indigo. It *feels* like something to taste feta cheese.

Somehow, these conscious perceptions are underpinned by neural activity—and the search for how, where, and why became a legitimate pursuit beginning in the 1990s. This shift in the zeitgeist inspired scientists to think about how aspects of conscious experience can be described, talked about, and measured. At first it seemed easy to assume that personal, subjective experience comes in only one flavor—but everyone soon realized that synesthesia provides a clear counterexample that could not be ignored. Thus, the new interest in personal experience suddenly made synesthesia a serious matter: Do people really have a different experience? How can it be measured? What does it mean for brain function? From the mechanical days of behaviorism, the acceptance of internal experience has catapulted synesthesia into the spotlight.

Together, new technologies and the resurgence of interest in consciousness have combined to make synesthesia a field of its own at the beginning of the twenty-first century. The aim of this book has been to synthesize the exciting progress of the field in one place. In this last chapter we turn the spotlight on where the field is going.

The Future of Synesthesia

Science has only begun to tap available technologies for studying synesthesia. We saw in the last chapter how fMRI demonstrates increased activity in color areas in the brain when a grapheme → color synesthete views a letter. The ground is wide open for future studies of this type. For example, one could use fMRI to see whether a grapheme → personality synesthete shows increased activation in parts of the brain associated with emotional salience and person familiarity (an area called retrosplenial cortex). These tests could physiologically verify that some people show different brain responses to letters and numbers than others. Other techniques can also be leveraged. For example, EEG recordings (using electrodes on the scalp) or galvanic skin resistance measurements (the technique used in lie detectors) could be used when a grapheme → personality synesthete views a number with an emotionally salient association. The results could be compared to the same subject's viewing emotionally weak numbers or punctuation and other symbols that lack emotional salience.

In the future dedicated researchers will launch forward–looking (prospective) studies, tracking data on large groups of infants as they develop. Doing so, we may be able to eventually pin down our speculation that synesthesia results from specific windows of time during which gene expression and environment interact in just the right manner to produce neural cross talk that sticks.

Along these lines, we expect our understanding of the genetics to grow exponentially in the near future. When cross-talking brain regions are sensory, the phenomenon of synesthesia easily reveals itself by its unusual perceptual manifestations. However, once we identify a gene or region, an important next question will be whether other people in the population possess the same mutation but express the gene in different parts of the brain. What would be the consequence of increased cross talk between brain regions that are not sensory—for example, between frontal areas involved in cognition or moral reasoning? What happens when areas involved in memory and planning express higher than normal interaction? Could this be the basis of increased creativity, intelligence, or madness? Our future understanding of the mechanism of synesthesia may shed light on mental, cognitive, and emotional talents or disorders.

As we press forward with new technologies, it will be critical to identify the many open questions and direct experiments toward them. We turn to several of these questions now.

How Much Consistency Is Expected?

For all forms of synesthesia discussed in this book, consistency testing has become the gold standard.[3] The idea is that a verifiable synesthete will always report the same synesthetic response to the same triggering stimulus, even when responding several months or years apart. It is easy to demonstrate that nonsynesthetes who are asked to fake it will fail to show good consistency.

Although consistency testing successfully discriminates synesthetes from controls, an open (if generally undiscussed) question remains: how much consistency should we expect? Is some amount of drift normal? To his surprise, David found upon reexamining Erica F that she changed from consistently labeling G green on three trials to consistently choosing brown on three trials on the retest. The time period between the two tests was just over six months, and when Erica was shown her old color choice, she expressed surprise and exclaimed, "I could have sworn my G has always been brown." We hope that by carefully quantifying other cases of color drift over time, researchers can hone our expectations of how much consistency to look for over long periods of time.

We also anticipate a drift for colored words whose colors are often modulated by word meaning. Recall, for example, Cassidy C's change in the color for "phthalocyanine" when he learned that it was the name of a blue–green pigment (as shown in figure 3.5). Children with colored alphabets are constantly exposed to new words; as they learn definitions and positive or negative associations seep in during a lifetime of learning, word color can change quite a bit, drifting from letter-dominated to meaning-dominated coloration. These considerations suggest there will eventually be a closer relationship between neuroscience, neurolinguistics, and synesthesia.

Drift over time is also quite apparent in number forms. Recall the morphing forms of Colleen Silva, who reports that the number line for her age in years changes in shape as she ages (see figure 5.10). This is also true for year forms, which evolve with passing time: for most synesthetes, the location of future years (say, 2017) are ill-defined, but when the year 2017 arrives, it will occupy a much more distinct position.

Why Are Grapheme → Color Associations Patchy?

Careful inspection of colored alphabets coupled with posttest interviewing makes it clear that many grapheme → color synesthetes do not have colors for all of their letters, or at least not all with the same strength of association. For example, the first synesthete in figure 3.3 has no color association for 9, whereas the second synesthete has no colors for I, O, and Q, and the third has none for C. This patchiness is not unusual. Even synesthetes who report their entire alphabet in color make some responses with high confidence while other letter–color pairings are weaker. This has opened the door for the neuroimaging studies that contrast the presentation of letters that trigger colors (or trigger them most strongly) against letters that do not trigger colors (or trigger them weakly)—all within the same subject.[4]

But what does this patchiness mean for our understanding of synesthesia? First, it means our basic theory of cross talk between areas is too simple. Despite the story suggested by fMRI results, it is not the case that *all* graphemes strongly activate the color areas—instead different graphemes' cross talk entails varying degrees of strength.

Generally, the observation of patchy associations supports the metaphor of a mountainous landscape, in which cross-area connections have a variety of strengths and therefore varying abilities to push activity over some threshold of consciousness (see figure 10.1). While the metaphoric model remains speculative, it gives a foothold for considering observations such as these.

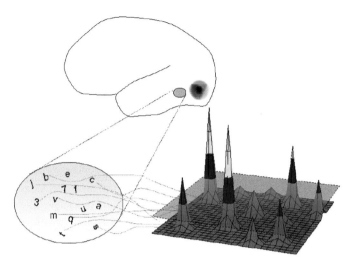

Figure 10.1

In this cartoon, neurons coding for graphemes and those coding for colors are connected with varying strengths. Thus, some graphemes are able to push activity in the color area above a threshold for consciousness (represented by the blue surface), while other activations remain below the level of detection.

Are the Types of Synesthesia Nonrandom?

Why is *color* the synesthetic property most often triggered by stimuli? Looking at table 2.1, it is clear that color accounts for most synesthetic experiences. But why are other equally plausible forms not often found, such as hearing a musical pitch when looking at a letter? The answer will probably not be found using an argument about proximity of brain structures because, for example, area AIT, a brain area involved in word meaning, is closer to the primary auditory cortex, which contains a map of pitches, than to V4, which contains a color map (see chapter 9). Thus, if close neighbors were more likely to be cross-talking than areas further apart, we would expect letters and words to trigger pitches, not colors.

These observations raise the possibility of an evolutionary aspect of synesthesia. Specifically, the distribution of sensory pairings in table 2.1 is not random. Instead, the skewed distribution may reflect evolutionary selection pressures for and against certain types. Imagine a synesthesia in which certain concepts caused auditory hallucination. Since sound has only a single channel, this could interfere with hearing actual environmental sounds. Vision, on the other hand, has many dimensions: think of shape, movement, location, contrast, and so on. Color is just one of these dimensions and is certainly not necessary (think of the population of color-blind people who get by just fine). Thus, a synesthesia that links to color vision is less likely to cause destructive interference than one linking into more critical functions such as touch or hearing. Correspondingly, the latter types are much rarer. The examination of the statistics of synesthesia types from an evolutionary point of view is just beginning but promises to eventually provide a framework for understanding many aspects of the phenomenon.

Why Is Synesthesia Unidirectional?

Synesthesia is typically unidirectional, meaning that while J may be blue, the color blue does not trigger the experience of J–ness.[5] When we colloquially talk about areas engaging in cross talk or being highly interconnected, it is important to emphasize that these connections are directional: they tend to be one-way streets. But why should the neural cross talk not go the other way? One possibility is that is *does* go both ways, but in a manner that lives below the threshold of awareness.

Recently Roi Cohen Kadosh and colleagues set out to test this assumption of unidirectionality. They found several number–color synesthetes, all of whom reported that colors do not evoke numbers in their daily life experience. The researchers presented them with a modified size-congruity task to examine whether colors triggered any subconscious aspect of numbers. To do this, they turned to the classic size congruity paradigm, in which the physical size of a digit and the numerical value vary independently. Although participants are instructed to attend to the numerical value and ignore the physical size, size is processed automatically and thus affects performance.[6] For example, when the pair 3 6 is presented, processing of the numerical information (e.g., "Which number is higher?") is facilitated if the numerically larger digit is also physically larger (e.g., if 6 is physically larger than 3). On the other hand, if the numerically smaller digit is physically larger (e.g., when 3 is physically larger than 6), interference occurs, and processing of the numerical information is slowed. In a modified version of this task, Cohen Kadosh found that synesthetic colors, as well, can implicitly evoke numerical magnitudes. For example, imagine that a synesthete experiences a small number (say, 1) in blue and a large number (say, 9) in green. When she is presented with a fairly congruent blue 3 and a green 6, she would be more likely to respond to a numerical judgment quickly and correctly. In contrast, when she is presented with a green 3 and a blue 6, she will be slower and less accurate. This is because the color appears to be unconsciously triggering the number's magnitude, which (as described above) interferes with the numerical judgment.

The simple finding that color can evoke numerical magnitude calls into question the assumption of unidirectionality—that is, that numbers trigger color—and forces the field to think carefully about what it expects to find at the level of microanatomy. Additional evidence like Cohen Kadosh's will suggest that unidirectional neural circuits may not be what we should be seeking.

Is Autism the Flip Side of Synesthesia?

As many people know from the movie *Rain Man*, autism is a genetic disorder that impairs social and communicative development while restricting interests and activities. Aside from these deficits, autism often goes hand in hand with savant skills in math, drawing, and music. Compared to normal controls, autistics also often excel on tests of visuospatial ability and rote memory. For example, they tend not to be fooled by certain visual illusions that routinely take in everyone else.

The pattern of deficits and skills in autism has led some researchers to propose that the cognitive style of autistics is biased toward local rather than global information processing. This position is termed "weak central coherence."[7,8] The gist is that autistics perform well when attention to details is required (i.e., piecemeal processing) but poorly when the forest must be grasped in favor of the trees. This observed difference is in accordance with autobiographical accounts of autism that often recount a fragmentation of perception.[9]

Consistent with this hypothesis, autism researcher Francesca Happé discovered that autistics do not perceive certain visual illusions.[10] For example, in the Ebbinghaus illusion a subject compares the size of two circles that are themselves surrounded by circles of different sizes (see figure 10.2).[11] Even though the two central circles are identical in size, normal brains are fooled into thinking the circle on the right is larger. The illusion depends entirely on the surrounding context—if the surrounding circles are taken

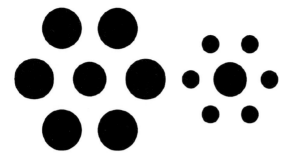

Figure 10.2
Although the two middle circles are of identical size, the one on the right seems larger to most people. The illusion is driven by local context cues.

away, the middle circles necessarily look identical. As predicted by weak central coherence, however, autistics are less likely to be fooled by the illusion. In terms of what we have discussed throughout this book, this fact loosely suggests that normal neural cross talk (in this case, between areas that tie items in with their surrounding context) is *reduced* in autistic brains.

Many similar findings support this view of autism. For example, when normal subjects are asked to count dots, the task is often helped by the global shape the dots make. Not so with autistics.[12] Similarly, autistics are less susceptible to visually induced motion.[13] For example, if all the stripes on the wallpaper in front of you suddenly moved to the right, you would involuntarily step to the right, thinking you had swayed to the left. A similar situation holds true for binding sight and sound. We discussed the McGurk effect in which the sight of moving lips influences what sound is heard. Autistics have a greatly reduced McGurk effect, again showing that their visual and auditory systems are not tied together as tightly as those of normal individuals.[14]

While the weak central coherence theory is still a subject of debate, it is enticing to consider its opposite nature from synesthesia. Neural cross talk is reduced in autism but increased in synesthesia. Typically, for any direction of change we find in nature, a change in the opposite direction also exists. Our view of the relationship between autism and synesthesia is at present speculative, and will certainly be refined by future data. However, we consider it worth mentioning as a starting framework for thinking about normal cross talk in the brain.

Does Synesthesia Negatively Influence Cognition?

Synesthetes generally love having their peculiar perceptions, even when at times they are unpleasant or lead to embarrassing situations. We have shown how synesthesia enhances life, not just by the pleasant affect it affords but also by usefully improving memory.

On the flip side, however, there are times when synesthesia can be burdensome or confusing—when, for example, percepts pile up to an overwhelming sensory overload or in the rare cases when bidirectional synesthesia leads to a perceptual feedback loop. There is also the possibility, just beginning to be explored in more depth, that a few subtle cognitive problems may accompany synesthesia, such as calculation, finger identification, right–left confusion, and sense of direction. We turn to these briefly.

A fair proportion of synesthetes (76%) claim to be poor at arithmetic despite having excellent memories for numbers. When formally tested, a minority of individuals reach the threshold for having a clinical impairment in the ability to calculate, called "dyscalculia." The difficulty is one of manipulating arithmetical symbols. A larger fraction exhibits more subtle problems such as cross-coding written or auditory numeric words into digits (e.g., seeing "five" then having to write down "5") or vice versa.[15]

As pointed out in chapter 5, the left angular gyrus in the parietal lobe is crucial for numerical aptitude. We have tested a few synesthetes, Michael Watson and Deni Simon among them, who also exhibited a striking angular gyrus sign called "finger agnosia." It is one of those peculiar clinical circumstances for which neurology is famous, consisting of the weird inability to tell one's fingers apart. Normally, the loss of finger identity would be utterly unimportant were it not for the fact that the failure so strongly localizes to the left angular gyrus and, thus, is a useful sign in clinical practice.

Because not everyone can readily name their fingers (e.g., "index," "ring," "middle"), testing is done by simply labeling the fingers one through ten. Individuals are then asked, "Show me your second finger" and so forth. Finger agnosia exists as the failure to associate a verbal label to one's fingers. An alternate method that does not require any verbal label relies on touch, proprioception, and an intact body schema. In this method, the subject closes his or her eyes and places the hands palms down on the table. The examiner then touches various combinations of two fingers while asking "how many" fingers are between the ones being touched. A surprising number of synesthetes make errors when doing this sensory task, thus implicating their angular gyrus.

Another shortcoming synesthetes as a group *tend* to show is right–left confusion, technically called "allochiria." So far, we have not found it in any one kind of synesthesia but rather across various types. One tests it by asking subjects to name or point to body parts, especially the examiner's hands held in a crisscross posture. This neurological finding similarly localizes to the vicinity of the angular gyrus, in this case the left TPO junction.

A related left parietal lobe test that some synesthetes also fail is the ability to recognize by touch alone graphemes that are written on the tip of their index finger. This failure is called "fingertip agraphesthesia." We have encountered synesthetes who score perfectly when letters or numerals are written on their left index finger (indicating an intact *right* parietal lobe). However, when the same task is done using their right index finger, the same subjects make mistakes such as calling a 9 a 0, or a 6 a 3. Such errors are again consistent with a *left* parietal anomaly in these individuals.

Last among synesthetes' possible cognitive deficits is a poor sense of direction. Again, this has been noted in a variety of synesthesia types rather than in any one form alone. Richard Cytowic initially found that 69% of forty-two subjects claimed to have a terrible sense of direction. Individuals went to unusual lengths to get around, even in cities laid out on a grid system such as New York or Washington, DC. For example, Rita Bush laments,

I have no sense of direction, which causes me distress. I cannot visualize where one location is in relation to another, even though they are very familiar places to me.

I can't tell you how relieved I felt when you told me this was comon [sic] with synesthesia. My husband says I'm the only person who tried to go through a turnstile in the wrong direction. If I'm driving and come to a dead-end street I can't get back to a parallel street and get terribly lost. I worked in the same office building for 10 years and still have trouble finding friends' offices.

Muriel Nolan makes a similar complaint:

I have no sense of direction. I cannot read a map. In driving to a new location
(which has been murder for me this Summer looking for a new job and going
to several interviews a week [*in various new locations*]) I must have very specific
directions. Most of the time, I have to "practice" a day in advance before going
some place new. Then I must write it down, very specifically in my own words
and descriptive phrases. Most people think I am an idiot because I cannot read
a map.

On visiting Richard Cytowic's hotel for an interview, Suzanne deM had
to twice get instructions on how to find the elevators despite their location
being obvious:

I have to have a map of every place I go to. I have absolutely no sense of direction,
even in cities I live in. Coming up out of the subway, I have to stop to figure out
what side of the street I'm on. I'm always getting turned around. I've lived in Boston
for eight years now and I still get lost. No one in my family has this problem. They
don't need maps.

Several issues appear to be involved here. The first is that a "good
memory" does not apply to all aspects of memory such as place finding.
In the face of boasting that they can easily recount conversations long
since past, interview others without benefit of taking notes, or recall the
location of reference material even down to the specific page of a book,
the prospect that some synesthetes suffer poor *spatial* memory is intriguing. It seems paradoxical to have a spatial impairment when synesthesia
so often intertwines with number forms and other concepts involving
spatial relationships.

Geographic knowledge and map reading are examples of spatial aptitude that depends on the right parietal region. When we walk through
our own homes, give directions to a stranger, mentally envision which
shops we can visit during our lunch break and what the most efficient
sequence would be, we are said to be using topographic knowledge or
a "cognitive map." The properties of cognitive maps should not be misunderstood to resemble those of ordinary cartographic maps. They are
not read and memorized but are *learned* by engaging the world. Whereas

visual perception has a single viewpoint, cognitive maps are built up by direct experience gained from multiple viewpoints.

The observation that synesthetes have uneven cognitive skills requires follow up with sufficiently large samples. So far, one such large-scale study of 192 adult lexical → color synesthetes confirms difficulties with math and with distinguishing right from left, as well as a poor sense of direction coupled with impaired map reading.[16]

Future studies will also aim to answer the question of whether synesthesia influences an individual's temperament—for example, with respect to whether the individual has an affinity for art and artistic expression. In *The Mind of a Mnemonist*, Luria pondered the influence of synesthesia on personality:[17]

Is it reasonable to think that the existence of an extraordinarily developed figurative memory, of synesthesia, has no effect on an individual's personality structure? Can a person who "sees" everything; who cannot understand a thing unless an impression of it "leaks" through all his sense organs; who must feel a telephone number on the tip of his tongue before he can remember it—can he possibly develop as others do? . . . Indeed one would be hard put to say which was more real for him: The world of imagination in which he lived or the world of reality in which he was but a temporary guest.

Whatever the final pros and cons of synesthesia, whatever its cognitive blessings and curses, most synesthetes evaluate it positively. Most say that, given a choice, they would never relinquish their synesthesia.

Reality Is Not One Size Fits All

It is a rule in science that nature reveals herself through exceptions. Accordingly, synesthesia is no mere curiosity. Rather, it is a window looking out on a broad swath of the mind, the brain, and our highly individualized view of what constitutes reality. All of us, synesthetes and nonsynesthetes alike, sense just a small fraction of outside reality. For example, we do not perceive the electromagnetic spectrum in its entirety, but merely a tiny slice of it. The rest of the spectrum—carrying TV shows, radio signals, and cell phone conversations—flows through our bodies with no awareness on our part. We are utterly blind to it. Our brains sample just a small bit of the surrounding physical world to construct reality. This is different from the commonsense view that eyes, ears, and fingers passively receive an "objective" physical world outside of ourselves.

Synesthesia, in its dozens of varieties, highlights the amazing differences in how individuals subjectively see the world, reminding us that each brain uniquely filters what it perceives in the first place. The world is more subjective than objective.

On the first page of this book we introduced one of our synesthetic subjects, Erica Borden, who has several varieties of synesthesia. When we first met her, she had no idea she experienced the world differently than her friend Aviva. Erica now appreciates her perceptions in a different light. She knows that an infinitesimally tiny genetic change has caused areas in her brain to have increased cross talk. She wonders: is this due to increased wiring? Decreased inhibition? Synaptic connections that got stuck in place? We do not yet know the answers to this, or what else the gene will mean for her cognition or personality. We also do not know what new tests Erica will take in a lab ten years from now: computerized tests that will accurately capture subjective experience, brain scans that will detail the microanatomy of her neural connections, and molecular screenings that will pinpoint the locations and time frames of the expression of her synesthesia genes. Time will tell.

In the meantime, Erica is growing up in a world where people will not think she is crazy or dismiss her as a liar, and where the science of the brain will be fueled by the treasures it discovers in her skull. And so, as Erica watches the weather patterns, experiences the taste of the chocolate covered raisins on her skin, and sees colorful sounds, she can draw on the knowledge of scientists as they draw on her experience, and everyone can learn more about brains and experience from her unusual experience of synesthesia.

Afterword

On a rain-soaked evening in 1937 or 1938, on a Paris sidewalk, I was tugging at my mother's hand to peer through a shop window at something that held a special fascination: a display whose details I do not remember but which was illuminated in a rich, flashing red. Like many small children, I would invent names for objects that particularly attracted me. There is an old Russian word for the color "red," which I had probably heard in some fable that had been read to me. The word contains the two Russian letters most difficult to render in English, so I shall not confuse the reader and instead limit myself to the essence: the word's phonemes became, in my mind, indelibly associated with that particular tint of red—a rich, luminous carmine—and the adjective generated a noun in my infant vocabulary, "*alochki* [ah-loch-ki]," denoting just that kind of miniature colored extravaganza. Red is also linked, in my mind, with the musical note "la," and had become so before I learned that, in English musical notation, this note is called "A." The color remains associated with the grapheme in other linguistic contexts as well and in the sundry shadings that vary with nuances of pronunciation from language to language. In music the association is not simply with the written name of the note but with its sound as well. And this goes somewhat further, into territory that requires a musical ear. The key in which a particular composition is written, played, or sung gives the piece an overall coloring. For example, Schubert's "*Doppelgänger*," when performed in E-flat minor, has a deep yellowish shade, while, if it is done in E minor, the hue approaches white. I have begun with samples from music because that is a domain to which I have devoted much of my life. Coloration, however, extends in my case to numbers, to the ensemble of an occurrence, to an individual, or to a train of thought. It can go further. I am not a religious person in the sense that

I do not endorse a liturgy or pray in formulae. Nonetheless, for a number of years, when there was something I devoutly wished—say, the well-being of a loved one—my yearning tended to be integral with the sense of a profound cavity, with a dark violet number 4 in its depths. The more distinct that image, the more likely I considered the fulfillment of my wish. And, in a more general sense, the more intently I have wished that that loved one should recover, or that something—say, an opera performance or a sporting event—would go well for me, the more the thought process has taken place against a background of a particular hue, generally in the red–violet part of the spectrum.

In this brief overview, as the reader will have noticed, I concentrate on personal and familial experience rather repeating the second-hand information of varying reliability that can be found in a literature that ranges from rigidly scientific texts to the yellow press. I shall continue in the same vein, except for an occasional digression. One example of the latter, amid the increasing mentions of synesthesia in today's press, concerns an item recently reported in *Seed* magazine, in an article entitled "The Most Beautiful Painting You've Ever Heard," and is about the experiences of an artist named Marcia Smilack, whose rare form of neurological mixing involves all of her senses:

> The sound of one female voice looks like a thin bending sheet of metal, and the sight of a certain fishing shack gives her a brief taste of Neapolitan ice cream—but her artistic leanings are shared by many other synesthetes. Scientists estimate that synesthesia is about seven times more common in poets, novelists, and artists than in the rest of the population. (Some of the most famous examples include artists David Hockney and Wassily Kandinsky and writer Vladimir Nabokov.)

A different, fictional instance that might be called dream synesthesia, or else poetic synesthesia, comes from a distant part of the chronological spectrum. It is an episode from Aleksandr Pushkin's verse drama *Boris Godunov*. I categorize the case as fictional because the plot of Pushkin's work differs radically from more authentic fact as it appears in the account of the historian Karamzin, on which Pushkin loosely based his work. Boris did not, in reality, murder young Dmitri the Pretender so as to mount the throne, while in the drama, and in Mussorgsky's opera, his hands and his conscience are incarnadine with the child's blood. The Orthodox Patriarch is brought before Boris to relate an ancient shepherd's incriminating tale that will precipitate the Tsar's fatal crisis. He is so old and ill, recounts the old man, that his dreams no longer communicate the visible and he can only dream sounds. An extraordinary characteristic of synesthesia, recently described by Daniel Levitin, a cognitive psychologist and specialist in the neuroscience of music at McGill University in Montreal, takes us literally in the opposite direction: from the grave to the cradle. Among the striking scientific trivia discovered by this polymath is the possibility that babies begin life with synesthesia. Normally, the cerebral cross talk gets sorted out as the individual develops, while subjects who remain under the influence of five-sense synesthesia either learn to discipline the exceptional combination and benefit by the total recall that it accords or are driven to distraction by the incessant input of superfluous information encumbering the brain.

My father has on occasion written of colored synesthesias in the more limited realm, familiar to him, of what he called colored hearing or colored vision, as elicited by letters of the alphabet. An early recollection of his concerns a set of building blocks he received as a gift from his mother—the kind that has a letter of the alphabet on every facet. "But the colors are all wrong!" he immediately complained after unwrapping the gift, because they did not correspond to the colors associated in his mind with the depicted letters. To digress: in the first part of his childhood my father was a mathematical prodigy who could perform complex operations with fantastically long strings of numbers. Could there be an analogy here with A. R. Luria's description of Shaseresevsky's "ability to remember limitless amounts" on the part of certain synesthetes? The young Nabokov fell ill, and when the fever broke, he had lost all semblance of this talent. It is interesting to conjecture whether this aberration was in any way associated with synesthesia. Or whether Nabokov's very serious passion for lepidopterology, and the wealth of color that entomology in general contained, was in some way reciprocal with his synesthesia. What I do know is that psychoneural phenomena intrigued him, be it Technicolor dreams, déjà vu, or colored hearing. He would have welcomed the advent of synesthesia in the scientific community and the serious research that was spearheaded by Richard Cytowic and taken up by younger colleagues such as David Eagleman. Curiously, I had received and begun reading Dr. Cytowic's *The Man Who Tasted Shapes* shortly before we both appeared on a BBC documentary titled "Orange Sherbet Kisses." Having been interviewed previously on various subjects, I had expected to be asked to talk freely about my personal synesthetic experiences. Instead I was handed a script, containing little that was new, to memorize—something that anyone, synesthete or not, could easily have done. What a pity that a program that might have been truly informative for the lay spectator could not toe a finer line rather than deteriorating into a reality show that culminated with a twangy American lady and the rock group whose very sight was her infallible route to orgasm.

Perhaps the most significant domain in which synesthesia may have affected Vladimir Nabokov was that of metaphor. When he describes an object, be it a chance item or an important prop, odds are that his description will have not only a touch of originality but also a color. Without burdening this foreword with fascinating but space-consuming lists of examples, let me cite only a few instances. For example, why, in the short play "The Grand-Dad," does the gallows—in a scene set during the French Revolution—need to be blue? Both of my parents were synesthetes. It was interesting to determine whether any of my colorations reflected a melding of the parental colors, but this was not the case except for a very muddy variant of the letter f. It has been shown that a synesthete's color associations tend to remain constant throughout his or her life. In our case, it was possible to determine this to a substantial degree. My father tested me when I was eight, and again when I was in my thirties. Those letters of the alphabet that were colored had retained their original hue. The same can be said for the constancy of coloration, or simply of aura, associated with a particular event or recollection.

Occasionally, unexpected manifestations of color occur in Vladimir Nabokov's writing, beginning at a relatively early age, but I have found it difficult to establish consistent color pathways in his case. I have just translated from Russian into English and Italian a very early story of his, which may in fact have been his first (at present I am awaiting the result of a color scan from the Library of Congress that may shed some light on a date that is practically illegible in the manuscript). The story's heroine, named Natasha, during a ramble through a deserted lakeside café with a suitor, imagines that a wind orchestra is playing "orange" music on the abandoned bandstand, and that music is colored "orange"—clear as day in the manuscript with no room for error. The head of the suitor himself is described as light blue, with a frequency that increases as the timid man's love becomes more evident. Synesthetic phenomena recur throughout Nabokov's work and are assigned to characters in two of his major English-language works, *Lolita* and *Ada*. This characteristic of my father's work has made me sensitive to its occurrence in the work of other writers as well. One can cite the great Russian poet Tyutchev, whom my father both trans-lated and recorded. For example, in Tyutchev's poem *Vchera v mechtakh obvorozhennykh* . . . (Last Night, amidst Charmed Reverie . . .), reference is made to a "*scarlet*, vibrant exclamation [*italics mine*]."

Many kinds and instances of presumed synesthesia, some convincing, some less so, have been described in literature and ascribed to noted poets, artists, and musicians, as well as sundry unremarkable individuals. Having familiarized myself with the types and categories of the phenomenon described by Richard Cytowic, David Eagleman, and others, I tend to con-serve the conviction that my synesthesia is in some ways unique or previ-ously undescribed in the documentation that I have seen.

First of all, if it were true that aging causes deterioration of "brain cells that function as chatty go-betweens," one might conclude that synesthe-sia, too, decreases with age. In my case, contemplated from the perspective of my seventy-three years, this appears not to be the case.

Furthermore, the considerable thought I have given to the implications and applications of synesthesia have yielded some interesting creative fruit. The problems inherent in the transformation of original, beautiful written language into its visual equivalent has long been a conundrum for the cineast and a seemingly insurmountable obstacle when approaching the filming of a Nabokov work, as well as works of others, such as Joyce, where language and imagery also play a crucial part when he is not being too wordy. That is why neither of the two *Lolitas*, while fine films in their own right, nor the other movies based on my father's works were as satisfying as they might have been. I am currently faced with the possibility of making a film of the novel *Ada*, perhaps the most complex and variegated of Nabokov's books. Rather than prune its text of its abundant poetic descriptions and reduce its rich language to the skeletal platitudes of the usual TV series, or accept the axiom that visual depiction of beautiful prose is impossible, my conception—which may in fact lead to a series—is a succession of images as perceived by Van, Ada, and the other characters, each with his own colors and variations of shape, each a distorting lens through which the delicately, deliberately disjointed components of the author's construct unite and realign themselves into a congruent whole. The palette is infinite, and originality is limited only by the imagination of the creator. One character may see another tinged with the aura corresponding to a particular emotion, or perhaps encircled by spikes suggesting antipathy, or perhaps tinted with an otherwordly blue, as we, through Natasha's eyes, tend to see Wolfe. As for the intensity of an orgasm, it may give birth not only to geometric aberrations in the mind but, for example, to a seemingly unending tunnel of pleasure through which one races in a crescendo of sensation toward the ultimate release. And a new kind of cinema, perhaps, thanks to an enlightened perception of synesthesia.

Dmitri Nabokov
Montreux, February 2008

Notes

Chapter 1

1. Broadcast August 31, 1993, "Book Talk" on WNYC, New York.

2. Bowers H, Bowers JE. 1961. *Arithmetical Excursions*. New York: Dover, pp. 244–247.

3. Correspondence to Dr. Cytowic, November 11, 1986.

4. Luria AR. 1968. *The Mind of a Mnemonist*. New York: Basic Books, p. 81.

5. His S-cones, which are sensitive to short wavelengths, are abnormal, making it difficult to distinguish blues and purples. The example comes from Ramachandran VS, Hubbard EM. 2001. Synaesthesia—A window into perception, thought and language. *Journal of Consciousness Studies* 8(12): 3–34, p. 24.

6. Ramachandran VS, Hubbard EM. 2001. Synaesthesia—A window into perception, thought and language. *Journal of Consciousness Studies* 8(12): 3–34, p. 2.

7. Galton F. 1880. Visualized numerals. *Nature* 22: 494–495.

8. Simner et al. 2005. Non—random associations of graphemes to colors in synaesthetic and normal populations. *Cognitive Neuropsychology* 22(8): 1069–1085.

9. Cytowic RE. 2002. *Synesthesia: A Union of the Senses*, 2nd ed. Cambridge: MIT Press, p. 55.

10. Baron-Cohen S, Burt L, Smith-Laittan F, et al. 1996. Synaesthesia: Prevalence and familiarity. *Perception* 25: 1073–1079.

11. Emrich HM, Schneider U, Zeidler M. 2000. *Welche Farbe hat der Montag?* Stuttgart: Hirzel Verlag.

12. Ramachandran VS, Hubbard EM. 2001. Synaesthesia—A window into perception, thought, and language. *Journal of Consciousness Studies* 8(12): 3–34.

13. Mattingley JB, Ward J (eds.). 2006. *Cognitive Neuroscience Perspectives on Synaesthesia* (Special Issue). *Cortex* 42: 129–320.

14. Harrison's observations agree with ours. Harrison J. 2001. *Synaesthesia: The Strangest Thing.* New York: Oxford University Press.

15. Galton F. 1883. *Enquiries into Human Faculty and Its Development.* London: Macmillian and Co.

16. Nabokov V. 1949. Portrait of my mother. *New Yorker* April 9, pp. 33–37. See also chapter 2 of *Speak, Memory: An Autobiography Revisited* (1966). New York: Dover (first published in 1951 as *Conclusive Evidence*).

17. As discussed in the BBC *Horizon* documentary "Orange Sherbet Kisses," broadcast December 13, 1994.

18. For example, English HB. 1923. Colored hearing. *Science* 57: 444; Riggs LA, Karwoski T. 1934. Synaesthesia. *British Journal of Psychology* 25: 29–41; Werner H. 1940. *Comparative Psychology of Mental Development.* New York: Harper.

19. Hall GS. 1883. The contents of children's minds. *Princeton Review* 11(4th serial): 249–272.

20. Marks LE. 1975. On colored–hearing synesthesia: Cross-modal translations of sensory dimensions. *Psychological Bulletin* 82(3): 303–331.

21. Simner J, Ward J, Lanz M, et al. 2005. Non-random associations of graphemes to colors in synaesthetic and normal populations. *Cognitive Neuropsychology* 22(8): 1069–1085; Simner J, Mulvenna C, Sagiv N, Tsakanikos E, Witherby SA, Fraser C, Scott K, Ward J. 2006. Synaesthesia: The prevalence of atypical cross-modal experiences. *Perception* 35(8): 1024–1033.

22. As discussed in the BBC *Horizon* documentary "Orange Sherbet Kisses," broadcast December 13, 1994.

23. Jordan DS. 1917. The color of letters (letter). *Science* 46(1187): 311–312; see also Riggs and Karwoski (1934), who present four children whose color associations changed over time.

24. Cytowic RE. 2002. *Synesthesia: A Union of the Senses,* 2nd ed. Cambridge: MIT Press, p. 292 (table 7.2).

25. For example, her 1919 "Music—Pink and Blue II."

26. Scriabin's music key–color correspondences are based on the regular circle of fifths, and he appears to have borrowed his color scheme from Madame Blavatsky of Theosophy fame. His *Prometheus, The Poem of Fire* (1911) calls for a *clavier á lumiéres*, a mute keyboard that controls the play of colored light in the form of beams, clouds, and other effects. For more about color organs, see Peacock K. 1988. Instruments to perform color–music: Two centuries of technological instrumentation. *Leonardo* 21: 397–406.

27. See "Concepts of Mind," pp. 25–54, and "Concepts of Neural Tissue," pp. 55–136, in Cytowic RE. 1996. *The Neurological Side of Neuropsychology*. Cambridge: MIT Press.

28. Mesulam M.-M. 2000. *Principles of Behavioral and Cognitive Neurology*, 2nd ed., pp. 1–120. New York: Oxford University Press.

29. Churchland P. 1979. *Scientific Realism and the Plasticity of Mind*. Cambridge: Cambridge University Press. See also Cytowic RE. 2003. The clinician's paradox: Believing those you must not trust. *Journal of Consciousness Studies* 10(9–10): 157–166.

30. Cytowic RE. 2003. The clinician's paradox: Believing those you must not trust. *Journal of Consciousness Studies* 10(9–10): 157–166, p. 160.

31. Cytowic RE. 1997, p. 24 in S Baron–Cohen, JE Harrison (eds.), *Synaesthesia: Classic and Contemporary Readings*. Oxford: Blackwell.

32. Cytowic RE. 2002. *Synesthesia: A Union of the Senses*, 2nd ed. Cambridge: MIT Press, pp. 67–69.

33. Nagel T. 1986. *The View from Nowhere*. New York: Oxford University Press.

Chapter 2

1. Available at http://home.comcast.net/~sean.day/html/the_synesthesia_list.html. Given how rapidly URLs change, check either http://Cytowic.net or www .eaglemanlab.net/synesthesia for up-to-date links to The Synesthesia List.

2. Simner J, Ward J, Lanz M, et al. 2005. Non-random associations of graphemes to colors in synaesthetic and normal populations. *Cognitive Neuropsychology* 22(8): 1069–1085; Simner J, Mulvenna C, Sagiv N, Tsakanikos E, Witherby SA, Fraser C, Scott K, Ward J. 2006. Synaesthesia: The prevalence of atypical cross-modal experiences. *Perception* 35(8): 1024–1033.

3. Suarez de Mendoza, 1890; Holden, 1891; Flournoy, T. (1893). Des Phénomènes de Synopsie, On the Phenomena of Synopsia, Charles Eggimann & Co., Genève (1893); Calkins, M.W. (1893). A Statistical Study of Pseudo-chromesthesia and of Mental-forms, *American Journal of Psychology* 5: 439–464; Bos, MC. (1929). Über echte und unechte audition coloreé. *Zeitschrift für Psychologie*; 111: 321–401; Wellek, Albert (1931). Zur Geschichte und Kritik der Synästhesie-Forschung. *Archiv für die gesamte Psychologie* 79, S.325–384; Kloos, Gerhard (1931). Synästhesien bei psychisch Abnormen. Eine Studie über das Wesen der Synästhesie und der synästhetischen Anlage. *Archiv für Psychiatrie und Nervenkrankheiten* 94, 418–469.

4. Galton first noted these in an 1880 *Nature* paper (21: 252–256) titled "Visualzed Numerals." These later appeared in the 1883 and 1907 editions of *Enquiries into Human Faculty and Its Development.*

5. Galton F. 1883. *Enquiries into Human Faculty and Its Development.* London: Macmillian and Co., pp. 80–81.

6. Feynman RP. 1988. *What Do You Care What Other People Think?* New York: Harper-Collins, p. 59.

7. Generally the letters learned first tend to be the ones in the child's name. They may know a few letters at age three but cannot map them to phonemes for a year. For both counting and alphabet sequences, children learn the rote routine way before they have any concept of what the letters or numbers map onto. See Wynn K. 1990. Children's understanding of counting. *Cognition* 36(2): 155–193.

8. For a good general review of the concepts and fallacies, see Baily DB, Bruer JT, Symons FJ, Lichtman JW. 2001. *Critical Thinking about Critical Periods.* Baltimore: Brookes Publishing Co.

9. Paulesu E, Harrison J, Baron-Cohen S, et al. 1995. The physiology of coulored hearing: A PET activation study of coulored-word synaesthesia. *Brain* 118: 661–676.

10. Patterson KE, Morton J. 1985. From orthography to phonology: An attempt at an old interpretation, pp. 335–359 in KE Patterson, JC Marshall, M Coltheart (eds.), *Surface Dyslexia.* Hillsdale, NJ: Lawrence Erlbaum.

11. Nabokov V. 1966. *Speak, Memory: An Autobiography Revisited.* New York: Dover (first published in 1951 as *Conclusive Evidence*).

12. Ward J, Simner J. 2003. Lexical–gustatory synaesthesia: Linguistic and conceptual factors. *Cognition* 89: 237–261.

13. Correspondence to Dr. Cytowic of May 1, 1987.

14. Correspondence to Dr. Cytowic of September 2, 1985.

15. Flournoy, T. 1893. Des Phénomènes de Synopsie, written at Geneva and Paris. Alcan.; Calkins MW. 1895. Synesthesia. *American Journal of Psychology* 7: 90–107.

16. Devereaux G. 1966. An unusual audio–motor synesthesia in an adolescent. *Psychiatric Quarterly* 40(3): 459–471.

17. Interview of October 19, 1984.

18. Starr F. 1893. Note on colored hearing. *American Journal of Psychology* 51: 416–418.

19. Stroop JR. 1935. Studies of interference in serial verbal reactions. *Journal of Experimental Psychology* 28: 643–662.

20. MacLeod CM. 1991. Half a century of research on the Stroop effect: An integrative review. *Psychological Bulletin* 109: 163–203.

21. Dixon MJ, Smilek D, Cudahy C, Merikle PM. 2000. Five plus two equals yellow: Mental arithmetic in people with synaesthesia is not coloured by visual experience. *Nature* 406(6794): 365.

22. Ward J, Huckstep B, Tsakanikos E. 2006. Sound–colour synaesthesia: To what extent does it use cross-modal mechanisms common to us all? *Cortex* 42: 264–280.

23. Edquist J, Rich AN, Brinkman C, Mattingley JB. 2006. Do synaesthetic colours act as unique features in visual search? *Cortex* 42: 222–231.

24. Palmeri TJ, Blake R, Marois R, et al. 2002. The perceptual reality of synesthetic colors. *Proceedings of the National Academy of Sciences, USA* 99: 4127–4131. See also Blake R, Palmeri T, Marois R, Kim C-O. 2004. On the perceptual reality of synesthesia, pp. 47–73 in LC Robertson, N Sagiv (eds.), *Synesthesia: Perspectives from Cognitive Neuroscience*. New York: Oxford University Press.

25. Paulsen HG, Laeng B. 2005. Pupillometry of grapheme–color synaesthesia. *Cortex* 42: 290–294.

26. The phrase was coined by Simon Baron-Cohen.

27. Cytowic RE. 2002. *Synesthesia: A Union of the Senses*, 2nd ed. Cambridge: MIT Press, pp. 84–98.

28. Galton F. 1883. *Enquiries into Human Faculty and Its Development*. London: Macmillian and Co., p. 107.

29. Simner JA, Ward JA, Lanz MA, Jansari AA, Noonan KA, et al. 2005. Non-random associations of graphemes to colours in synaesthetic and non-synaeasthetic populations. *Cognitive Neuropsychology* 22(8): 1069–1085.

30. Luria AR. 1968. *The Mind of a Mnemonist*. New York: Basic Books, p. 28.

31. Haber RN, Haber RB. 1964. Eidetic imagery. I: Frequency. *Perceptual and Motor Skills* 19: 131–138; Haber RN. 1969. Eidetic images. *Scientific American* 220: 36–44; Haber RN. 1979. Twenty years of haunting eidetic imagery: Where's the ghost? *Behavioral and Brain Sciences* 2: 583–629.

32. Jaensch ER. 1930. *Eidetic Imagery and Typological Methods of Investigation*, 2nd ed. (O Oeser, trans.). New York: Harcourt Brace.

33. Cytowic RE. 2002. *Synesthesia: A Union of the Senses*, 2nd ed. Cambridge: MIT Press, pp. 103–110.

34. Haber RN, Haber RB. 1964. Eidetic imagery. I: Frequency. *Perceptual and Motor Skills* 19: 131–138.

35. Stromeyer CF, Psotka J. 1970. The detailed texture of eidetic images. *Nature* 225: 346–349.

36. Dann KT. 1999. *Bright Colors Falsely Seen: Synesthesia and the Search for Transcendental Knowledge*. New Haven: Yale University Press; Johnson DB. 1985. *Worlds in Regression: Some Novels of Vladimir Nabokov*. Ann Arbor: Ardis.

37. Nabokov V. 1969. *Ada or Ardor: A Family Chronicle*. New York: McGraw-Hill, p. 584.

38. Nabokov V. 1962. *Pale Fire*. New York: Putnam, p. 34.

39. Correspondence of October 10, 1985.

40. Correspondence to Dr. Cytowic of May 6, 2001.

41. Posting to The Synesthesia List of 2/14/06.

42. Klüver H. 1966. *Mescal and Mechanisms of Hallucinations*. Chicago: University of Chicago Press, p. 22.

43. Siegel RK. 1977. Hallucinations. *Scientific American* 237(4): 132–140; Siegel RK, Jarvik ME. 1975. Drug-induced hallucinations in animals and man, pp. 81–162 in RK Siegel, LJ West (eds.), *Hallucinations: Behavior, Experience, and Theory*. New York: Wiley.

44. Horowitz MJ. 1975. Hallucinations: An information processing approach, in RK Siegel, LJ West (eds.), *Hallucinations: Behavior, Experience and Theory*. New York: Wiley.

45. Cytowic, RE. 2002. Simple synesthesia and deafferentation, pp. 114–120 in *Synesthesia: A Union of the Senses*, 2nd ed. Cambridge: MIT Press.

46. McKellar P. 1957. *Imagination and Thinking*. New York: Basic Books.

47. Downey J. 1911. A case of colored gestation. *American Journal of Psychology* 22: 528–539.

48. Siegel RK, Jarvik ME. 1975. Drug-induced hallucinations in animals and man, pp. 81–162 in RK Siegel, LJ West (eds.), *Hallucinations: Behavior, Experience, and Theory*. New York: Wiley.

Chapter 3

1. Jordan DS. 1917. The colors of letters. *Science* 46(1187): 311–312.

2. Baron-Cohen S, Burt L, Smith-Laittan F, et al. 1996. Synaesthesia: Prevalence and familiality. *Perception* 25(9): 1073–1079.

3. Eagleman DM, et al. 2006. A standardized test for synesthesia.

4. Simner J, Glover L, Mowat A. 2006. Linguistic determinants of word colouring in grapheme–colour synaesthesia. *Cortex* 42: 281–289.

5. Dixon MJ, Smilek D, Merikle PM. 2004. Not all synaesthetes are created equal: Projector versus associator synaesthetes. *Cognitive, Affective and Behavioral Neuroscience* 4(3): 335–343.

6. Maurer D. 1997. Neonatal synesthesia: Implications for the processing of speech and faces, pp. 224–242 in S Baron-Cohen, JE Harrison (eds.), *Synaesthesia: Classic and Contemporary Readings*. Cambridge: Blackwell Publishers; also see Ramachandran and Hubbard, 2001.

7. Nunn JA, Gregory LJ, Brammer M, et al. 2002. Functional magnetic resonance imaging of synesthesia: Activation of V4/V8 by spoken words. *Nature Neuroscience* 5: 371–375.

8. Ramachandran, VS & Hubbard, EM. (2001). Psychophysical investigations into the neural basis of synaestehsia. Proceedings of the Royal Society of London B. 268: 979–983.

9. Myles KM, Dixon MJ, et al. 2003. Seeing double: The role of meaning in alphanumeric–colour synaesthesia. *Brain and Cognition* 53(2): 342–345.

10. Dixon MJ, Smilek D, Cudahy C, Merikle PM. 2000. Five plus two equals yellow. *Nature* 406(6794): 365.

11. Smilek D, Dixon MJ, Cudahy C, Merikle PM. 2002. Concept driven color experiences in digit–color synesthesia. *Brain and Cognition* 48(2–3): 570–573.

12. Hubbard EM, Manohar S, Ramachandran VS. 2006. Contrast affects the strength of synesthetic colors. *Cortex* 42: 184–194.

13. Baron-Cohen S, Harrison J, et al. 1993. Coloured speech perception: Is synaesthesia what happens when modularity breaks down? *Perception* 22(4): 419–426.

14. Day S. 2005. Some demographic and socio-cultural aspects of synesthesia, pp. 11–33 in LC Robertson, N Sagiv (eds.), *Synesthesia: Perspectives from Cognitive Neuroscience*. New York: Oxford University Press.

15. Simner J et al. 2005. Non-random associations of graphemes to colors in synaesthetic and normal populations. *Cognitive Neuropsychology* 22(8): 1069–1085.

16. Berlin B, Kay P. 1969. *Basic Color Terms: Their Universality and Evolution*. Berkeley: University of California Press.

17. Shanon B. 1982. Colour associates to semantic linear orders. *Psychological Research* 44: 75–83.

18. Witthoft N, Winawer J. 2006. Synesthetic colors determined by having colored refrigerator magnets in childhood. *Cortex* 42: 175–183.

19. Rich AN, Bradshaw JL, Mattingley JB. 2005. A systematic large-scale study of synaesthesia: Implications for the role of early experience in lexical–color associations. *Cognition* 98: 53–84.

20. Simner J, Glover L, Mowat A. 2006. Linguistic determinants of word colouring in grapheme–colour synaesthesia. *Cortex* 42: 281–289; Simner J et al. 2005. Non-random associations of graphemes to colors in synaesthetic and normal populations. *Cognitive Neuropsychology* 22(8): 1069–1085.

21. Smilek D, Dixon MJ, Cudahy C, Merikle PM. 2002. Synesthetic color experiences influence memory. *Psychological Science* 13(6): 548–552.

22. Simner J, Holenstein E. 2007. Ordinal linguistic personification as a variant of synesthesia. *Journal of Cognitive Neuroscience* 19(4): 694–703; Simner J, Hubbard EM. 2006. Variants of synesthesia interact in cognitive tasks: Evidence for implicit associations and late connectivity in cross-talk theories. *Neuroscience* 143(3): 805–814.

23. Day S. 2005. Some demographic and socio-cultural aspects of synaesthesia. In LC Robertson, N Sagiv (eds.), *Synaesthesia: Perspectives from Cognitive Neuroscience.* Oxford: Oxford University Press.

24. Noam Sagiv, Olufemi Olu-Lafe, Maina Amin, and Jamie Ward (2006). Grapheme personification: A profile. *UK Synaesthesia Association 2nd Annual Meeting* (April 22–23). London, UK.

25. Brumbaugh RS. 1981. *ThePhilosophers of Greece.* Albany: State University of New York Press.

26. Calkins MW. 1895. Synaesthesia. *American Journal of Psychology* 7, 90–107.

27. Note that the French scientist T. Flournoy addressed personified graphemes in 1893; it is unclear whether Calkins would have been aware of his writing.

28. Calkins MW. 1895. Synaesthesia. *American Journal of Psychology* 7, 90–107, p. 100.

29. Simner, J, Holenstein E. (2007). Ordinal linguistic personification as a variant of synesthesia. *Journal of Cognitive Neuroscience.* 19(4):694–703.

30. Noam Sagiv, Olufemi Olu-Lafe, Maina Amin, and Jamie Ward (2006). Grapheme personification: A profile. *UK Synaesthesia Association 2nd Annual Meeting.* April 22–23. London, UK.

Chapter 4

1. Cytowic, interview of October 21, 1983.

2. English H. 1923. Colored hearing. *Science* 57: 444; Helson H. 1933. A child's spontaneous reports of imagery. *American Journal of Psychology* 45: 360–361.

3. Riggs L, Karwoski T. 1934. Synesthesia. *British Journal of Psychology* 25: 29–41; Simpson R, Quinn M, Ausubel D. 1956. Synesthesia in children: Associations of colors with pure tone frequencies. *Journal of Genetic Psychology* 89: 95–103.

4. Whitchurch AK. 1922. Synesthesia in a child of three and a half years. *American Journal of Psychology* 33: 302–303.

5. Cytowic, correspondence of July 6, 2001.

6. Ziegler MJ. 1930. Tone shapes: A novel type of synesthesia. *Journal of General Psychology* 3: 227–287.

7. Eagleman, personal communication with Laurel Smith, October 18, 2005.

8. Quoted from an anonymous article in the *Neuen Berliner Musikzeitung* (August 29, 1895); also quoted in Mahling F. 1926. Das Problem der "Audition colorée": Eine historische-kritische Untersuchung. *Archiv für Gesamte Psychologie* 57: 165–301.

9. Cytowic RE. 2002. *Synesthesia: A Union of the Senses*. 2nd ed. Cambridge: MIT Press, pp. 208–312.

10. Samuel C. 1976. *Conversations with Olivier Messiaen* (F Aprahamian, trans.). London: Stainer and Bell, p. 93.

11. Messiaen O. 1977. *Aux canyons des etoiles*. Liner notes, Erato STU70974/975 (recording). Paris: Alphonse Leduc.

12. Bernard JW. 1986. Messiaen's synaesthesia: The correspondence between color and sound structure in his music. *Music Perception* 4(1): 41–68.

13. Messiaen O. 1944. *The Technique of My Musical Language*, vol. 1 (J Satterfield, trans.). Paris: Alphonse Leduc, p. 51.

14. Samuel C. 1976. *Conversations with Olivier Messiaen* (F Aprahamian, trans.). London: Stainer and Bell; Goléa A. 1960. *Rencontres avec Olivier Messiaen*. Paris: Julliard.

15. Amoore JE. 1977. Specific anosmia and the concept of primary odors. *Chemical Senses and Flavor* 2: 267–281.

16. Profitta J, Bidder H. 1988. Perfect pitch. *Journal of Musical Genetics* 29(4): 763–771.

17. Schlaug G, Jancke L, Huang Y, Steinmetz H. 1995. In vivo evidence of structural brain asymmetry in musicians. *Science* 267: 699–701.

18. Über das Geistige in der Kunst (1912; Bern, 1952); trans. and ed. Kenneth C Lindsay and Peter Vergo, under the title *On the Spiritual in Art. Kandinsky: Complete Writings on Art*, 2 vols. (Boston, 1982).

19. She was featured in the BBC *Horizon* documentary, "Orange Sherbet Kisses," broadcast December 13, 1994.

20. Marks LE. 1974. On associations of light and sound: The mediation of brightness, pitch, and loudness. *American Journal of Psychology* 87: 173–188; Marks LE, Hammeal RJ, Bornstein MH. 1987. Perceiving similarity and comprehending metaphor. *Monographs of the Society for Research in Child Development* 52: 1–93; Marks LE. 2004. Cross-modal interactions in speeded classification, pp. 85–106 in G Calvert, C Spence, BE Stein (eds.), *Handbook of Multisensory Processes*. Cambridge: MIT Press.

21. Hubbardd TL. 1996. Synesthesia-like mappings of lightness, pitch, and melodic interval. *American Journal of Psychology* 109: 219–238.

22. Ward J, Huckstep B, Tsakanikos E. 2005. Sound–colour synaesthesia: To what extent does it use cross-modal mechanisms common to us all? *Cortex* 42: 264–280.

23. Rizzo MR, Esslinger PJ. 1989. Colored hearing synesthesia: Investigation of neural factors in a single subject. *Neurology* 39: 781–784.

24. Maurer D, Mondlach CJ. 2005. Neonatal synesthesia: A reevaluation, pp. 193–213 in LC Robertson, N Sagiv (eds.), *Synesthesia: Perspectives from Cognitive Science*. New York: Oxford University Press; Cytowic RE. 2002. *Synesthesia: A Union of the Senses*, 2nd ed., pp. 271–293.

25. Stein BE, Meredith MA. 1993. *The Merging of the Senses*. Cambridge: MIT Press.

26. Vroomen J, De Gelder B. 2004. Perceptual effects of cross-modal stimulation: Ventriloquism and the freezing phenomenon, pp. 141–150 in G Calvert, C Spence, BE Stein (eds.), *Handbook of Multisensory Processes*. Cambridge: MIT Press.

27. McGurk G, MacDonald J. 1976. Hearing lips and seeing voices. *Nature* 264: 746–748.

28. Schwartz J, Robert-Ribes J, Escudier JP. 1998, p. 319 in R Campbell, B Dodd, DK Burnham (eds.), *Hearing by Eye*. East Sussex: Hove.

29. Shams L, Kamitani Y, Shimojo S. 2000. Illusions: What you see is what you hear. *Nature* 408(6814): 788.

30. Gebhard and Mowbray, 1959; Shipley, 1964; Welch, Duttonhurt, and Warren, 1986.

31. Eagleman DM. 2001. Visual illusions and neurobiology. *Nature Reviews Neuroscience* 2(12): 920–926.

32. Loe PR, Benevento LA. 1969. Auditory-visual interaction in single units in the orbito-insular cortex of the cat. *Electroencephalography and Clinical Neurophysiology* 26: 395–398; Benevento LA, Fallon J, Davis BJ, Rezak M. 1977. Auditory-visual

interaction in single cells in the cortex of the superior temporal sulcus and the orbital frontal cortex of the macaque monkey. *Experimental Neurology* 57: 849–872; Meredith MA, Nemitz JW, Stein BE. 1987. Determinants of multisensory integration in superior colliculus neurons. I: Temporal factors. *Journal of Neuroscience* 7: 3215–3229; Eagleman, 2008 *Cortex* (in Press).

33. Calvert GA et al. 1997. Activation of auditory cortex during silent lipreading. *Science* 276: 593–596.

34. Macaluso E, Frith CD, Driver J. 2000. Modulation of human visual cortex by crossmodal spatial attention. *Science* 289: 1206–1208.

35. de Gelder B, Bocker KB, Tuomainen J. et al. 1999. The combined perception of emotion from voice and face: Early interaction revealed by human electric brain responses. *Neuroscience Letter* 260: 133–136.

36. Lewkowicz D, Turkewitz G. 1980. Cross-modal equivalence in infancy: Auditory-visual intensity matching. *Developmental Psychology* 16: 597–607.

37. Marks LE. 2004. Cross-modal interactions in speeded classification, pp. 85–106 in G Calvert, C Spence, BE Stein (eds.), *Handbook of Multisensory Processes*. Cambridge: MIT Press.

38. Wallace MT. 2004. The development of multisensory integration, pp. 625–642 in G Calvert, C Spence, BE Stein (eds.), *Handbook of Multisensory Processes*. Cambridge: MIT Press.

39. Eagleman DM. In press.*Ten Unsolved Questions of Neuroscience*. New York: Oxford University Press.

40. Neville HJ. 1995. Developmental specificity in neurocognitive development in humans. In M Gazzaniga (ed.), *The Cognitive Neurosciences*. Cambridge: MIT Press.

41. Falchier A, Clavagnier S, Barone P, Kennedy H. 2002. Anatomical evidence of multimodal integration in primate striate cortex. *Journal of Neuroscience* 22: 5749–5759.

42. Innocenti et al., 1988.

43. Lickliter R, Bahrick LE. 2004. Perceptual development and the origins of multi-sensory responsiveness, pp. 643–654 in G Calvert, C Spence, BE Stein (eds.), *Handbook of Multisensory Processes*. Cambridge: MIT Press.

Chapter 5

1. Galton F. 1880. Visualized numerals. *Nature* 22: 494–495; Smilek D, Callejas A, Merikle P, Dixon M. 2006. Ovals of time: Space–time synesthesia. *Consciousness and Cognition* 16(2): 507–519.

2. Wheeler RH, Cutsforth TD. 1921. The number forms of a blind subject. *American Journal of Psychology* 32: 21–25.

3. Piazza M, Pinel P, Dehaene S. 2006. Objective correlates of a peculiar subjective experience: A single-case study of number-form synaesthesia. *Cognitive Neuropsychology* 23(8): 1081–1082.

4. Dehaene S, Molko N, Cohen L, Wilson A. 2004. Arithmetic and the brain. *Current Opinion in Neurobiology* 14: 218–224; Dehaene S. 2001. Précis of the number sense. *Mind & Language* 16: 16–36; Dehaene S, Dehaene-Lambertz G, Cohen L. 1998. Abstract representations of numbers in the animal and human brain. *Trends in Neuroscience* 21: 355–361.

5. Sagiv N, Simner J, Collins J, Butterworth B, Ward, J. 2006. What is the relationship between synaesthesia and visuo-spatial number forms? *Cognition* 101(1): 114–128.

6. Shanon B. 1982. Color associates to semantic linear orders. *Psychological Research* 44: 75–83 (4.5% figure); Seron X, Peseenti M, Noel M-P, et al. 1992. Images of numbers or "when 98 is upper left and 6 is sky blue." *Cognition* 44: 159–196 (12% figure).

7. See Piazza et al., 2006; Sagiv et al., 2006b; Smilek et al., 2006.

8. McKelvie SJ, Rohrberg MM. 1978. Individual differences in reported visual imagery and cognitive performance. *Perceptual and Motor Skills* 46(2): 451–458.

9. Dehaene S, Bossini S, Giraux P. 1993. The mental representation of parity and numerical magnitude. *Journal of Experimental Psychology: General* 122: 371–396; Fias W, Brysbaert M, Geypens F, D'ydewalle G. 1996. The importance of magnitude information in numerical processing: Evidence from the SNARC effect. *Mathematical Cognition* 2: 95–110; Hubbard EM, Piazza M, Pinel P, Dehaene S. 2005. Interactions between number and space in parietal cortex. *Nature Reviews Neuroscience* 6(6): 435–448.

10. Hubbard EM, Piazza M, Pinel P, Dehaene S. 2005. Interactions between number and space in parietal cortex. *Nature Reviews Neuroscience* 6(6): 435–448.

11. Plodowski A, Swainson R, Jackson GM, Rorden C, Jackson SR. 2003. Mental representation of number in different numerical forms. *Current Biology* 13(23): 2045–2050.

12. See Hubbard et al. 2005.

13. McTaggart JME. 1908. The unreality of time. *Mind* 18: 457–484; Russell B. 1915. On the experience of time. *The Monist* 25: 212–233; Callender C. 2000. Shedding light on time. *Philosophy of Science* 67 (Proceedings): S587–S599.

14. For more on the philosophy of time, see McTaggart's B–properties; also see the papers of Craig Callender. Tenseless philosophers say that events in time are relational, just like "to the north of" is relational—that is, it highlights the relationships between things.

Chapter 6

1. Most people are unfamiliar with the fifth basic taste, termed umami or "meaty taste," prototypically represented by monosodium glutamate. The prototypical stimuli for the other four are sucrose (sweet), sodium chloride (salty), citric acid (sour), and quinine (bitter).

2. Pritchard TC, Macaluso DA, Eslinger PJ. 1999. Taste perception in patients with insular cortex lesions. *Behavioral Neuroscience* 113: 663–671.

3. Royet JP, Koenig O, Gregoire MC, et al. 1999. Functional anatomy of perceptual and semantic processing for odors. *Journal of Cognitive Science* 11(1): 94–109.

4. In a keynote speech at the United Kingdom Synaesthesia Association, Charles Spence suggested that it might not be reasonable to talk about independent senses of taste and smell but rather to refer to the composite "flavor" sense.

5. Stevenson RJ, Boakes RA. 2004, pp. 69–84 in G Calvert, C Spence, BE Stein (eds.), *Handbook of Multisensory Processes*. Cambridge: MIT Press.

6. Dravnieks A. 1985. *Atlas of Odor Character Profiles*. ASTM Data Series DS61. Philadelphia: AASTM.

7. Baeyens F, Eelen P, van den Bergh O, Crombez G. 1990. Flavor–flavor and color–flavor conditioning in humans. *Learning and Motivation* 21: 434–455.

8. Stevenson RJ, Boakes RA. 2004, pp. 69–84 in G Calvert, C Spence, BE Stein (eds.), *Handbook of Multisensory Processes*. Cambridge: MIT Press.

9. Interview of May 9, 1981.

10. Liu H, Hockenberry M, Selker T. 2005. http://web.media.mit.edu/~hugo/publications/papers/SIGGRAPH2005-SynaestheticRecipes.pdf.

11. Interview of March 1981.

12. For statistical results, see Cytowic RE. 2002. *Synesthesia: A Union of the Senses*, 2nd ed. Cambridge: MIT Press, pp. 91–97.

13. Cytowic RE. 2002. *Synesthesia: A Union of the Senses*, 2nd ed. Cambridge: MIT Press, pp. 86–98.

14. Harrison JE. 2001. *Synaesthesia: The Strangest Thing*. New York: Oxford University Press, pp. 170–174.

15. Doty RL, Shaman P, Dann M. 1984. Development of the University of Pennsylvania Smell Identification Test: A standardized microencapsulated test of olfactory function. *Physiology & Behavior* (Monograph) 32: 489–502. It is now the most widely used olfactory test in the world. Its research first showed that smell loss is one of the first signs of Alzheimer's disease, Parkinson's disease, and several other neurodegenerative disorders.

16. Cytowic RE. 2002. *Synesthesia: A Union of the Senses*, 2nd ed. Cambridge: MIT Press, pp. 138–144.

17. For detailed data, see Cytowic RE. 2002. *Synesthesia: A Union of the Senses*, 2nd ed. Cambridge: MIT Press, pp. 143–144.

18. For details, see Cytowic RE. 2002. *Synesthesia: A Union of the Senses*, 2nd ed. Cambridge: MIT Press, pp. 133–167.

19. See Sean Day and Dr. Eagleman in the Discovery Channel documentary *One Step Beyond*. http://youtube.com/watch?v=DvwTSEwVBfc.

20. Downey JE. 1911. A case of colored gustation. *American Journal of Psychology* 22: 528–539.

21. Luria AR. 1968. *The Mind of a Mnemonist*. New York: Basic Books, p. 82.

22. Luria AR. 1968. *The Mind of a Mnemonist*. New York: Basic Books, p. 23.

23. Luria AR. 1968. *The Mind of a Mnemonist*. New York: Basic Books, p. 134.

24. Luria AR. 1968. *The Mind of a Mnemonist*. New York: Basic Books, p. 82.

25. Schultze E. 1912. Krankhafter Wandertrieb, räumlich beschränkte Taubheit für bestimmte Töne und "tertiare" Empfindungen bei einem Psychopathen. *Zeitschrift für die gesamte Neurologie und Psychiatrie* 10: 399.

26. Beeli G, Esslen M, Jancke L. 2005. Synaesthesia: When coloured sounds taste sweet. *Nature* 434(7029): 38.

27. Ward J, Simner J, Auyeung V. 2005. A comparison of lexical–gustatory and grapheme–colour synaesthesia. *Cognitive Neuropsychology* 22(1): 28–41.

28. Ward J, Simner J. 2003. Lexical–gustatory synaesthesia: Linguistic and conceptual factors. *Cognition* 89: 237–261.

29. Broadcast of October 8, 2004.

30. Luria AR. 1968. *The Mind of a Mnemonist*. New York: Basic Books, p. 82.

31. Pierce AH. 1907. Gustatory audition: A hitherto undescribed variety of synaesthesia. *American Journal of Psychology* 18: 341–352.

32. Van Orden GC. 1987. A rows is a rose: Spelling, sound, and reading. *Memory and Cognition* 14: 371–386.

33. Gray JA, Chopping S, Nunn J, et al. 2002. Implication of synaesthesia for functionalism. *Journal of Consciousness Studies* 9: 5–31.

34. Ward J, Collins J, Auyeung V. 2003. Word–taste synaesthesia is an automatic and perceptual phenomenon. *Journal of Cognitive Neuroscience* 15(suppl): 51.

35. Hubbard EH, Ramachandran VS. 2005. Individual differences among grapheme–color synesthetes: Brain–behavior correlations. *Neuron* 45: 1–11.

36. These areas include the insula, in a deep nucleus called the claustrum, and in the parietal operculum, as well as an orbitofrontal taste-sensitive region on the bottom surface of the brain that is also sensitive to colors, smells, and edibility judgments. Royet JP, Hudry J, Zald JH, et al. 2001. Functional neuroanatomy of different olfactory judgments. *Neuroimage* 13: 506–519; Rolls ET, Bayliss LL. 1994. Gustatory, olfactory and visual convergence within the primate orbitofrontal cortex. *Journal of Neuroscience* 14: 5437–5452.

Chapter 7

1. Correspondence of April 2, 1986.

2. Posting to The Synesthesia List, September 2, 2005.

3. Riggs LA, Karwoski T. 1934. Synaesthesia. *British Journal of Psychology* 25: 29–41.

4. Posting to The Synesthesia List, September 6, 2005.

5. Posting to The Synesthesia List, September 1, 2005.

6. Correspondence of July 25, 2005. Her synesthesias are ([emotion → color, shape] + [pain → color] + [music → shaped touch]).

7. Ward J. 2004. Emotionally mediated synaesthesia. *Cognitive Neuropsychology* 21(7): 761–772.

8. Ramachandran VS, Hubbard EM. 2001. Psychophysical investigations into the neural basis of synaesthesia. *Proceedings of the Royal Society of London B* 268: 979–983.

9. D'Andrade R, Egan M. 1975. The colors of emotion. *American Ethnologist* 1: 49–63.

10. Shah NJ, Marshall JC, Safiris O, et al. 2001. The neural correlates of person familiarity: A functional MRI imaging study with clinical applications. *Brain* 124: 804–815; Maddock RJ, Buonocore MH. 1997. Activation of the left posterior cingulate gyrus by the auditory presentation of threat related words: An fMRI study. *Psychiatry Research* 75: 1–14.

11. Maddock JR. 1999. The retrosplenial cortex and emotion: New insights from functional neuroimaging of the human brain. *Trends in Neuroscience* 22: 310–316.

12. Weiss PH, Shah NJ, Toni I, Zilles K, Fink GR. 2001. Associating colours with people: A case of chromatic–lexical synaesthesia. *Cortex* 37: 750–753.

13. Orgasm triggers activity in the autonomic nervous system (both sympathetic and parasympathetic), with heavy activity in the limbic system of the brain. See Komisaruk BR, Whipple B. 2005. Functional MRI of the brain during orgasm in women. *Annual Review of Sex Research* 16: 62–86.

14. Account of May 9, 1981.

15. Ramachandran VS, Hubbard EM. 2001. Psychophysical investigations into the neural basis of synaesthesia. *Proceedings of the Royal Society of London B* 268: 979–983.

16. Swinkels WAM, Kuyk J, van Dyck J, Spinhoven PH. 2005. Psychiatric comorbidity in epilepsy. *Epilepsy and Behavior* 7: 37–50.

17. Gloor P, Olivier A, Quesney LF, et al. 1982. The role of the limbic system in experiential phenomena of temporal lobe epilepsy. *Annals of Neurology* 12: 129–144.

18. Schomer DL, O'Coonnor M, Spiers P, et al. 2000. Temporolimbic epilepsy and behavior, pp. 377–388 in M-M Mesulam (ed.), *Principles of Behavioral and Cognitive Neurology*. New York: Oxford University Press.

19. Persinger MA. 1983. Religious and mystical experiences as artifacts of temporal lobe function: A general hypothesis. *Perceptual and Motor Skills* 57: 1255–1262; Persinger MA. 1989. The "visitor" experience and the personality: The temporal lobe factor. *Archaeus* 5: 157–171.

20. Gloor P. 1972. Temporal lobe epilepsy: Its possible contribution to the understanding of the significance of the amygdala and its interaction with neocortical–temporal mechanisms, pp. 423–457 in BE Eleftheriou (ed.), *The Neurobiology of the Amygdala*. New York: Plenum Press; Gloor P. 1986. Role of the human limbic system in perception, memory, and affect: Lesions from temporal lobe epilepsy, pp. 159–169 in BK Doane, KE Livingston (eds.), *The Limbic System: Functional Organization and Clinical Disorders*. New York: Raven Press.

21. Cytowic RE. 1996. *The Neurological Side of Neuropsychology*. Cambridge: MIT Press, pp. 402–404.

22. Ramachandran VS, Hirstein WS, Armel KC, et al. 1997. The neural basis of religious experience. *Society for Neuroscience Abstracts* 23: 1316.

23. Bear D. 1979. Temporal lobe epilepsy: A syndrome of sensory limbic hyperconnectionism. *Cortex* 15: 357–384.

24. The agent doing the explaining is likely the left-brain interpreter; see Gazzaniga MS, Eliassen JC, Nisenson L, Wessinger CM, Baynes KB. 1996. Collaboration between

the hemispheres of a callosotomy patient: Emerging right hemisphere speech and the left brain interpreter. *Brain* 119: 1255–1262. Also see Cytowic RE. 2008. *My Auto-Neurography: A Memoir of Intellect, Emotion, and Detachment.* In press, chapter 2.

Chapter 8

1. Ramachandran VS, Hubbard EM. 2001. Synaesthesia—A window into perception, thought, and language. *Journal of Consciousness Studies* 8(12): 3–34.

2. Marks LE. 1989. On cross-modal similarity: The perceptual structure of pitch, loudness, and brightness. *Journal of Experimental Psychology: Human Perception and Performance* 15: 586–602. See also Simpson L, Quinn M, Ausubel DT. 1956. Synesthesia in children: Association of color with pure tone frequencies. *The Journal of Genetic Psychology* 89: 95–103.

3. In addition to food coloring's changing the perceived taste of wine, it can even change the brain responses in a measurable way; see Osterbauer et al. 2005. Color of scents: Chromatic stimuli modulate odor responses in the human brain. *Journal of Neurophysiology* 93(6): 3434–3441.

4. See also Seitz JA. 2005. The neural, evolutionary, developmental, and bodily basis of metaphor. *New Ideas in Psychology* 23: 74–95.

5. Kohler W. 1929/1947. *Gestalt Psychology*, 2nd ed. New York: Liveright; Ramachandran and Hubbard called our attention to this work in their 2001 paper, Synesthesia—A window into perception, thoughts, and language. *Journal of Consciousness Studies* 8(12): 3–34.

6. Many of the examples are from G Lakoff, MH Johnson, 1980, *Metaphors We Live By*, Chicago: University of Chicago Press, and my discussion paraphrases theirs closely.

7. Marks LE, Hammeal RJ, Bornstein MH. 1987. Perceiving similarity and comprehending metaphor. *Monographs of the Society for Research in Child Development* 52: 1–93; Marks LE, Bornstein MH. 1987. Sensory similarities: Classes, characteristics, and cognitive consequences, pp. 49–65, in RE Haskell (ed.), *Cognition and Symbolic Structures: The Psychology of Metaphoric Transformation.* Norwood, NJ: Ablex.

8. Morgan GA, Goodson FE, Jones T. 1975. Age differences in the associations between felt temperatures and color choices. *American Journal of Psychology* 88(1): 125–130.

9. The psychologist Charles Osgood proposed there was a natural human tendency to think in terms of opposites. Osgood CE. 1952. The nature and measurement of meaning. *Psychological Bulletin* 49: 197–237.

10. Marks LE. 1974. On associations of light and sound: The mediation of brightness, pitch, and loudness. *American Journal of Psychology* 87: 173–188.

11. Marks LE. 1989. On cross-modal similarity: The perceptual structure of pitch, loudness, and brightness. *Journal of Experimental Psychology: Human Perception and Performance* 15: 586–602.

12. Day S. 1996. Synesthesia and synesthetic metaphors. *PSYCHE: An Interdisciplinary Journal of Research on Consciousness* 2(32). http://psyche.cs.monash.edu.au/v2/psyche-2-32-day.html

13. Bruner J. 1964. The course of cognitive growth. *American Psychologist* 19: 1–15.

14. Vygotsky L. 1965. *Thought and Language*. Cambridge: MIT Press.

15. James W. 1890. *The Principles of Psychology*. New York: Dover.

16. Meltzoff A, Moore M. 1992. Early imitation within a functional framework. *Infant Behavior and Development* 15: 479–505.

17. Tzourio-Mazoyer N, De Schonen S, Crivello F, et al. 2002. Neural correlates of woman face processing by 2-month-old infants. *Neuroimage* 15: 454–461.

18. Tellegen A, Atkinson G. 1978. Openness to absorbing and self altering experiences ("absorption"): A trait related to hypnotic susceptibility. *Journal of Abnormal Psychology* 83: 268–277.

19. Unpublished dissertation.

20. Domino G. 1989. Synesthesia and creativity in fine art students: An empirical look. *Creativity Research Journal* 2(1–2): 17–29; Ternaux JP. 2003. Synesthesia: A multimodal combination of senses. *Leonardo* 36: 321–322.

21. Interview with Dr. Cytowic on NBC's "Sightings."

22. Samuel C. 1976. *Conversations with Olivier Messaien* (F Aprahamian, trans.). London: Stainer and Bell, p. 125.

23. Samuel C. 1976. *Conversations with Olivier Messaien* (F Aprahamian, trans.). London: Stainer and Bell, p. 91.

24. Samuel C. 1976. *Conversations with Olivier Messaien* (F Aprahamian, trans.). London: Stainer and Bell.

25. Interview on "The Infinite Mind," National Public Radio, broadcast of January 12, 2005. Available at http://www.cafepress.com/lcmedia/350315.

26. Correspondence of August 10, 1981.

27. Duchting HJ. 1997. *Painting Music*. New York: Prestel, pp. 17, 65.

28. Dann K. 1998. *Bright Colors, Falsely Seen: Synesthesia and the Search for Transcendental Knowledge*. New Haven: Yale University Press.

29. Translation for the author by American poet Edwin Honig.

30. Berlin B, Kay P. 1969. *Basic Color Terms*. Berkeley and Los Angeles: University of California Press.

31. Myers CS. 1914–1915. Two cases of synaesthesia. *British Journal of Psychology* 7: 112–117.

32. We are indebted to musicologist Jörg Jewanski for the term.

33. Blavatsky HP. 1888/1999. *The Secret Doctrine: The Synthesis of Science, Religion and Philosophy*. Pasadena: Theosophical University Press.

34. Plummer HC. 1915. Color music—A new art created with the aid of science. The color organ used in Scriabin's symphony "Prometheus." *Scientific American* (April 10); Sullivan JWN. 1914. An organ on which color compositions are played: The new art of color music and its mechanism. *Scientific American* (February 21).

35. Fischinger O. 1949. "True Creation," Knokke-le-Zoute Film Festival notes, reprinted in W Moritz, 2004. *Optical Poetry: The Life and Work of Oskar Fischinger*. London: John Libbey, p. 192.

36. Disney W, quoted in W Moritz, 2004. *Optical Poetry: The Life and Work of Oskar Fischinger*. London: John Libbey, p. 84.

37. Individuals with a double dose of the sickle cell gene are severely ill. However, individuals who carry only one copy of the gene benefit by being resistant to malaria.

38. Hunt HT. 2005. Synaesthesia, metaphor, and consciousness. *Journal of Consciousness Studies* 12(12): 26–45.

39. Wild T, Kuiken D, Schopflocher D. 1995. The role of absorption in experiential involvement. *Journal of Personality and Social Psychology* 69: 569–579.

40. Root-Bernstein R, Root-Bernstein M. 1999. *Sparks of Genius: The Thirteen Thinking Tools of the World's Most Creative People*. Boston: Houghton Mifflin.

41. Daily A, Martindale C, Borkum J. 1997. Creativity: Synesthesia and physiognomic perception. *Creativity Research Journal* 10: 1–8.

42. Ramachandran VS. 2004. *A Brief Tour of Human Consciousness*. New York: Pi Press, p. 74.

43. Ramachandran and Hubbard, 2001.

44. Geschwind N. 1964, p. 155 in CJJM Stuart (ed.), *Monograph Series on Language and Linguistics, No. 17*. Washington, DC: Georgetown University.

45. Popper KR, Eccles JC. 1977. *The Self and Its Brain*. New York: Springer Verlag, p. 469.

46. Flechsig P. 1901. *Lancet* 2: 1027; Yakovlev P. 1962. Morpholological criteria of growth and maturation of the nervous system in man. *Research Publications of the Association for Nervous and Mental Disorders* 39: 3–46; Yakovlev PI, Lecours AR. 1967. The myelogenetic cycles of regional maturation of the brain, pp. 3–70 in A Minkowski (ed.), *Regional Development of the Brain in Early Life*. Oxford: Blackwell.

47. Pepperberg I. 1999. *The Alex Studies: Cognitive and Communicative Abilities of Grey Parrots*. Cambridge: Harvard University Press; Herman L, Richards D, Wolz J. 1984. Comprehension of sentences by bottle nosed dolphins. *Cognition* 16: 129–219.

48. Geschwind N. 1965. Disconnection syndromes in animals and man, Part I (p. 275). *Brain* 88: 237–294.

49. The phrase comes from Hans-Lucas Teuber. 1961. Sensory deprivation, sensory suppression and agnosia: Notes for neurologic theory. *Journal of Nervous and Mental Diseases* 132: 32–40.

50. Domino G. 1989. Synesthesia and creativity in fine art students: An empirical look. *Creativity Research Journal* 2(1–2): 17–29.

51. Raven's matrices is a nonverbal intelligence test that correlates well with the standard IQ test, the Wechsler Adult Intelligence Scale—Revised, and similar measures of verbal and performance IQ. Raven JC. 1956. *Guide to Using the Raven's Progressive Matrices*. London: HE Lewis.

52. Torrance JP. 1966. *Thinking Creatively with Words*. Princeton: Personnel Press.

53. Steen CJ. 2001. Visions shared: A firsthand look into synesthesia and art. *Leonardo* 34: 203–208; Ternaux JP. 2003. Synesthesia: A multimodal combination of senses. *Leonardo* 36: 321–322; Berman G. 1999. Synesthesia and the arts. *Leonardo* 32: 15–22.

Chapter 9

1. Eagleman, DM. 2007. Ten Unsolved Mysteries of the Brain. *Discover Magazine*. August 2007 issue.

2. Ramachandran VS, Hubbard EM. 2001. Psychophysical investigations into the neural basis of synaesthesia. *Proceedings of the Royal Society of London B* 268(1470): 979–983.

3. Edelman G. 1992. *Bright Air, Brilliant Fire*. New York: Basic Books; Zeki S. 1993. *A Vision of the Brain*. Oxford: Blackwell; Choo CW. 1998. *Information Management for the Intelligent Organization*. Medford, NJ: Information Today.

4. Eagleman DM. 2001. Visual illusions and neurobiology. *Nature Reviews Neuroscience* 2(12): 920–926.

5. McGurk H, MacDonald J. 1976. Hearing lips and seeing voices. *Nature* 264: 746–748; Schwartz J, Robert–Ribes J, Escudier JP. 1998, p. 319 in R Campbell, B Dodd, DK Burnham (eds.), *Hearing by Eye*. East Sussex: Hove; Van Wassenhove V, Grant KW, Poeppel D. 2007. Temporal window of integration in auditory–visual speech perception. *Neuropsychologia* 45(3): 598–607.

6. Calvert GA et al. 1997. Activation of auditory cortex during silent lipreading. *Science* 276: 593–596.

7. de Gelder B, Bocker KB, Tuomainen J, Hensen M, Vroomen J. 1999. The combined perception of emotion from voice and face: Early interaction revealed by human electric brain responses. *Neuroscience Letters* 260: 133–136.

8. Macaluso E, Frith CD, Driver J. 2000. Modulation of human visual cortex by crossmodal spatial attention. *Science* 289: 1206–1208.

9. Kennedy H, Batardiere A, Dehay C, Barone P. 1997. Synesthesia: Implications for developmental neurobiology, pp. 243–258 in S Baron-Cohen, JE Harrison (eds.), *Synaesthesia: Classic and Contemporary Readings*. Cambridge, Massachusetts: Blackwell.

10. Armel KC, Ramachandran VS. 1999. Acquired synesthesia in retinitis pigmentosa. *Neurocase* 5(4): 293–296.

11. Pascual-Leone A, Amedi A, Fregni F, Merabet LB. 2005. The plastic human brain cortex. *Annual Review of Neuroscience* 28: 377–401.

12. The first such demonstrations were presented in Sadato N, Pascual-Leone A, Grafman J, et al., 1996, 1998.

13. Lenay C, Gapenne O, Hanneton S, Marque C, Genouel C. 2003. Sensory substitution: Limits and perspectives, in *Touching for Knowing, Cognitive Psychology of Haptic Manual Perception*. Amsterdam/Philadelphia: John Benjamins, pp. 275–292.

14. Bach-y Rita P, Collins CC, Saunders F, White B, Scadden L. 1969. Vision substitution by tactile image projection. *Nature* 221: 963–964; Bach-y-Rita P. 2004. Tactile sensory substitution studies. *Annals of the New York Academy of Sciences* 1013: 83–91.

15. Cytowic RE. 1993. *The Man Who Tasted Shapes*. New York: Putnam.

16. Nunn JA, Gregory LJ, Brammer M, Williams SCR, Parslow DM, Morgan MJ, Morris RG, Bullmore ET, Baron-Cohen S, Gray JA. 2002. Functional magnetic resonance imaging of synesthesia: Activation of V4/V8 by spoken words. *Nature Neuroscience* 5: 371–375.

17. Unexpectedly, this group of synesthetes failed to activate their left V4 when actually looking at colors. That is, the participation of left V4 in synesthetic color experience renders it unavailable for ordinary color perception. In other words, synesthesia appears to hijack an existing brain function in this case. Alternatively, asymmetrical activation only on the left may have to do with the linguistic nature of the stimulus, language usually being a left-brain function. In general, these issues must be interpreted cautiously, because only a small sample of seven subjects was used.

18. Howard RJ et al. 1998. The functional anatomy of imagining and perceiving colour. *NeuroReport* 9: 1019–1023.

19. Paulesu E et al. 1995. The physiology of coloured-hearing: A PET activation study of colour–word synaesthesia. *Brain* 118: 661–676.

20. Rich AN, Mattingley JB. 2002. Anomalous perception in synaesthesia: A cognitive neuroscience perspective. *Nature Reviews Neuroscience* 3(1): 43–52.

21. Hubbard EM, Arman AC, Ramachandran VS, Boynton GM. 2005. Individual differences among grapheme–color synesthetes: Brain–behavior correlations. *Neuron* 45(6): 975–985.

22. These technical reasons include the spatial and temporal resolution of fMRI. More importantly, fMRI only detects changes in blood flow to a region. Often, there are important changes in neural activity with no corresponding changes in blood flow, making these signals invisible on brain scans.

23. Hubbard EM, Manohar S, Ramachandran VS. 2006. Contrast affects the strength of synesthetic colors. *Cortex* 42: 184–194.

24. Cohen L et al. 2000. The visual word form area: Spatial and temporal characterization of an initial stage of reading in normal subjects and posterior split-brain patients. *Brain* 123: 291–307.

25. Cohen L, Dehaene S. 2004. Specialization within the ventral stream: The case for the visual word form area. *NeuroImage* 22: 466–476.

26. Petersen SE, Fox PT, Snyder AZ, Raichle ME. 1990. Activation of extrastriate and frontal cortical areas by visual words and word-like stimuli. *Science* 249: 1041–1044.

27. Mechelli A et al. 2000. Differential effects of word length and visual contrast in the fusiform and lingual gyri during reading. *Proceedings of the Royal Society of London B* 267: 1909–1913.

28. Avidan G et al. 2002. Contrast sensitivity in human visual areas and its relationship to object recognition. *Journal of Neurophysiology* 87: 3102–3116.

29. Dixon MJ, Smilek D, Merikle PM. 2004. Not all synaesthetes are created equal: Projector vs. associator synaesthetes. *Cognitive, Affective and Behavioral Neuroscience* 4: 335–343.

30. Cohen et al., 2000.

31. Ramachandran VS, Hubbard EM. 2003. The phenomenology of synaesthesia. *Journal of Consciousness Studies* 10: 49–57.

32. Witthoft N, Winawer J. 2006. Synesthetic colors determined by having colored refrigerator magnets in childhood. *Cortex* 42: 175–183.

33. Hubbard EM, Piazza M, Pinel P, Dehaene S. 2005. Interactions between numbers and space in parietal cortex. *Nature Reviews Neuroscience* 6(6): 435–448.

34. Maurer, 1997; Ramachandran and Hubbard, 2001.

35. S Baron-Cohen et al. 1993. Coloured speech perception: Is synaesthesia what happens when modularity breaks down? *Perception* 22: 419–426.

36. Maurer D. 1997. Neonatal synaesthesia: Implications for the processing of speech and faces. In S Baron-Cohen, JE Harrison (eds.), *Synaesthesia: Classic and Contemporary Readings*. Cambridge, Massachusetts: Blackwell, pp. 224–242.

37. Maurer D. 1997. Neonatal synesthesia: Implications for the processing of speech and faces, pp. 224–242 in S Baron-Cohen, JE Harrison (eds.), *Synaesthesia: Classic and Contemporary Readings*. Cambridge: Blackwell.

38. Of course, this may reflect the difficulty in measuring synesthesia in infants.

39. Grossenbacher PG. 1997. Perception and sensory information in synesthetic experience. In S Baron-Cohen, JE Harrison (eds.), *Synaesthesia: Classic and Contemporary Readings*. Cambridge, Massachusetts: Blackwell, pp. 148–172.

40. Grossenbacher PG, Lovelace CT. 2001. Mechanisms of synesthesia: Cognitive and physiological constraints. *Trends in Cognitive Sciences* 5(1): 36–41.

41. Grossenbacher, 1997.

42. Grossenbacher and Lovelace, 2001.

43. Eagleman DM. 2009. *Plasticity: How the brain rewires itself on the fly*. New York: Oxford University Press.

44. Purpura DP. 1956a. Electrophysiological analysis of psychotogenic drug action. I: Effect of lysergic acid diethylamide (LSD) on specific afferent systems in the cat. *Archives of Neurology and Psychiatry* 75: 122–131; Purpura DP. 1956b. Electrophysiological analysis of psychotogenic drug action. II: General nature of lysergic acid diethylamide (LSD) action on central synapses. *Archives of Neurology and Psychiatry* 75: 132–143; Purpura DP. 1957. Experimental analysis of the inhibitory action of LSD on cortical dendritic activity. *Annals of the New York Academy of Sciences* 66: 515–536.

45. Renkel M. 1957. Pharmacodynamics of LSD and mescaline. *Journal of Nervous and Mental Diseases* 125: 424–427.

46. Cytowic RE. 2002. *Synesthesia: A Union of the Senses*, 2nd ed. Cambridge: MIT Press, p. 102.

47. Heron W. 1957. The pathology of boredom. *Scientific American* 196: 52–56.

48. Cytowic RE. 2002. *Synesthesia: A Union of the Senses*, 2nd ed. Cambridge: MIT Press, pp. 111–113.

49. Brust and Behrens, 1977.

50. Ramachandran VS, Armel C, 1999; discussed in Ramachandran VS, Hubbard EM, Butcher PA. 2004. Synesthesia, cross-activation, and the foundations of neuro-epistemology, pp. 867–883 in G Calvert, C Spence, BE Stein (eds.), *Handbook of Multisensory Processes*. Cambridge: MIT Press.

51. Miller and Crosby, 1979.

52. Jacobs L, Karpik A, Bozian D, Gøthgen S. 1981. Auditory–visual synesthesia: Sound-induced photisms. *Archives of Neurology* 38(4): 211–216; Cytowic RE. 2002, pp. 114–120.

53. Jacobs et al., 1981.

54. Lepore F. 1990. Spontaneous visual phenomena with visual loss: 104 patients with lesions of retinal and neural afferent pathways. *Neurology* 40: 444–447.

55. Vike J, Jabbari B, Maitland CG. 1984. Auditory–visual synesthesia: Report of a case with intact visual pathways. *Archives of Neurology* 41: 680–681.

56. Walsh R. 2005. Can synaesthesia be cultivated: Indications from surveys of meditators. *Journal of Consciousness Studies* 12(4–5): 5–17. Note that the synesthetes in this study were not rigorously verified.

57. Cytowic RE. 1993. *The Man Who Tasted Shapes*. New York: Putnam, p. 166.

58. Jiyu-Kennett PTNH. 1990. The scripture of great wisdom, in *The Litergy of the Order of Buddhist Contemplatives for the Laity*. Mt. Shasta, CA: Shasta Abbey Press, pp. 73–74.

59. Jacome and Gumnit, 1979.

60. Dudycha GJ, Dudycha MM. 1935. A case of synesthesia: Visual-pain and visual-audition. *Journal of Abnormal and Social Psychology* 30: 57–69.

61. Cytowic RE. 2002. *Synesthesia: A Union of the Senses*, 2nd ed. Cambridge: MIT Press, p. 124.

62. Hausser-Hauw C, Bancaud J. 1987. Gustatory hallucinations in epileptic seizures: Electrophysiological, clinical and anatomical correlates. *Brain* 110(Pt. 2): 339–359.

63. Galton F. 1883. *Enquiries into Human Faculty and Its Development.* London: Macmillian and Co.

64. Ward J, Simner J. 2005. Is synaesthesia an X-linked trait with lethality in males? *Perception* 34: 611–623.

65. Eagleman DM, Nelson S, Sarma SK. 2007. The neuroscience, behavior, and genetics of synesthesia. Presentation, Society for Neuroscience.

66. Smilek et al. 2002. Synaesthesia: A case study of discordant monozygotic twins. *Neurocase* 8: 338–342.

67. Thanks to Ed Hubbard for good discussion on this topic; see also Corballis MC. 1997. The genetics and evolution of handedness. *Psychological Review* 105: 714–777.

68. See the Simner et al. 2005 population studies.

69. Bailey MES, Johnson KJ. 1997. Synaesthesia: Is a genetic analysis feasible? pp. 182–207 in S Baron-Cohen, JE Harrison (eds.), *Synaesthesia: Classic and Contemporary Readings.* Cambridge: Blackwell.

70. Pascual-Leone A, Amedi A, Fregni F, Merabet LB. 2005. The plastic human brain cortex. *Annual Review of Neuroscience* 28: 377–401.

71. Witthoft and Winawer, 2006.

72. Cytowic RE. 1996. The standard hierarchical model, pp. 55–60 in *The Neurological Side of Neuropsychology.* Cambridge: MIT Press.

73. Shuler MG, Bear MF. 2006. Reward timing in the primary visual cortex. *Science* 311(5767): 1606–1609.

74. Toth LJ, Assad JA. 2002. Dynamic coding of behaviourally relevant stimuli in parietal cortex. *Nature* 415(6868): 165–168.

Chapter 10

1. Eagleman DM et al. 2007. A standardized test battery for the study of Synesthesia. *Journal of Neuroscience Methods* 159: 139–145.

2. Eagleman DM. 2005. Obituary: Francis H. C. Crick (1916–2004). *Vision Research* 45: 391–393.

3. Asher et al., 2006; Baron-Cohen and Harrison, 1997; Jordan, 1917; Eagleman et al., 2007.

4. Sperling JM, Prvulovic D, Linden DEJ, et al. 2006. Neuronal correlates of colour–graphemic syaesthesia: A fMRI study. *Cortex* 42: 295–303; Weiss PH, Zilles K, Fink GR. 2005. When visual perception causes feeling: Enhanced cross-modal processing

in grapheme–color synesthesia. *NeuroImage* 28: 859–868; Rouw R, Scholte HS. 2007. Increased structural connectivity in grapheme–color synesthesia. *Nature Neuroscience* 10: 792–797.

5. Mills CB, Boteler EH, Oliver GK. 1999. Digit synaesthesia: A case study using a Stroop-type test. *Cognitive Neuropsychology* 16: 181–191.

6. Henik A, Tzelgov J. 1982. Is three greater than five: The relation between physical and semantic size in comparison tasks. *Memory and Cognition* 10: 389–395.

7. Frith U. 1989. *Autism: Explaining the Enigma.* Oxford: Blackwell.

8. Happé F. 1999. Autism: Cognitive deficit or cognitive style? *Trends in Cognitive Sciences* 3(6): 216–222.

9. Gerland G. 1997. *A Real Person: Life on the Outside* (J. Tate, trans.). London: Souvenir Press.

10. Happé, FGE. 1996. Studying weak central coherence at low levels: Children with autism do not succumb to visual illusions: A research note. *Journal of Child Psychology and Psychiatry* 37: 873–877.

11. Eagleman DM. 2001. Visual illusions and neurobiology. *Nature Reviews Neuroscience* 2(12): 920–926.

12. Jarrold C, Russell J. 1997. Counting abilities in autism: Possible implications for central coherence theory. *Journal of Autism and Developmental Disorders* 27: 25–37.

13. Gepner B et al. 1995. Postural effects of motion vision in young autistic children. *NeuroReport* 6: 1211–1214.

14. de Gelder B, Vroomen J, Van der Heide L. 1991. Face recognition and lip-reading in autism. *European Journal of Cognitive Psychology* 3: 69–86.

15. Cytowic RE. 2002. *Synesthesia: A Union of the Senses*, 2nd ed. Cambridge: MIT Press, pp. 147–153.

16. Rich AN, Bradshaw JL, Mattingley JB. 2005. A systematic, large-scale study of synaesthesia: Implications for the role of early experience in lexical–colour associations. *Cognition* 98(1): 53–84.

17. Luria AR. 1968. *The Mind of a Mnenomist.* New York: Basic Books, pp. 150, 159.

Bibliography

Amoore, JE. (1977). Specific anosmia and the concept of primary odors. *Chemical Senses and Flavor* 2: 267–281.

Armel, KC, Ramachandran, VS. (1999). Acquired synesthesia in retinitis pigmentosa. *Neurocase* 5(4): 293–296.

Asher, J, Aitken, MRF, Farooqi, N, et al. (2006). Diagnosing and phenotyping visual synaesthesia: A preliminary evaluation of the revised test of genuineness (TOG–R). *Cortex* 42: 137–146.

Avidan, G, et al. (2002). Contrast sensitivity in human visual areas and its relationship to object recognition. *Journal of Neurophysiology* 87: 3102–3116.

Bach-y Rita, P, Collins, CC, Saunders, F, White, B, Scadden, L. (1969). Vision substitution by tactile image projection. *Nature* 221: 963–964.

Bach-y-Rita, P. (2004). Tactile sensory substitution studies. *Annals of the New York Academy of Sciences* 1013: 83–91.

Baeyens, F, Eelen, P, van den Bergh, O, Crombez, G. (1990). Flavor–flavor and color–flavor conditioning in humans. *Learning and Motivation* 21: 434–455.

Bailey, MES, Johnson, KJ. (1997). Synaesthesia: Is a genetic analysis feasible? pp. 182–207 in S Baron-Cohen, JE Harrison (eds.), *Synaesthesia: Classic and Contemporary Readings*. Oxford: Blackwell.

Baily, DB, Bruer, JT, Symons, FJ, Lichtman, JW. (2001). *Critical Thinking about Critical Periods*. Baltimore: Brookes.

Baron-Cohen, S, Harrison, J, et al. (1993). Coloured speech perception: Is synaesthesia what happens when modularity breaks down? *Perception* 22(4): 419–426.

Baron-Cohen, S, Burt, L, Smith-Laittan, F, et al. (1996). Synaesthesia: Prevalence and familiality. *Perception* 25(9): 1073–1079.

Baron-Cohen, S, Harrison, JE (eds.). (1997). *Synaesthesia: Classic and Contemporary Readings*. Oxford: Blackwell, pp. 224–242.

Bear, D. (1979). Temporal lobe epilepsy: A syndrome of sensory limbic hyperconnectionism. *Cortex* 15: 357–384.

Beeli, G, Esslen, M, Jäncke, L. (2005). When coloured sounds taste sweet. *Nature* 434: 38.

Benevento, LA, Fallon, J, Davis, BJ, Rezak, M. (1977). Auditory–visual interaction in single cells in the cortex of the superior temporal sulcus and the orbital frontal cortex of the macaque monkey. *Experimental Neurology* 57: 849–872.

Berlin, B, Kay, P. (1969). *Basic Color Terms: Their Universality and Evolution*. Berkeley: University of California Press.

Berman, G. (1999). Synesthesia and the arts. *Leonardo* 32: 15–22.

Bernard, JW. (1986). Messiaen's synaesthesia: The correspondence between color and sound structure in his music. *Music Perception* 4(1): 41–68.

Blake, R, Palmeri, T, Marois, R, Kim, C-O. (2004). On the perceptual reality of synesthesia, pp. 47–73 in LC Robertson, N Sagiv (eds.), *Synesthesia: Perspectives from Cognitive Neuroscience*. New York: Oxford University Press.

Blavatsky, HP. (1888/1999). *The Secret Doctrine: The Synthesis of Science, Religion and Philosophy*. Pasedena CA: Theosophical University Press.

Bos, MC. (1929). Über echte und unechte audition coloreé. *Zeitschrift für Psychologie* 111: 321–401.

Bowers, H, Bowers, JE. (1961). *Arithmetical Excursions*. New York: Dover, pp. 244–247.

Brumbaugh, RS. (1981). *The Philosophers of Greece*. Albany: State University of New York Press.

Bruner, J. (1964). The course of cognitive growth. *American Psychologist* 19: 1–15.

Brust, JCM, Behrens, MM. (1977). Release hallucinations as the major symptoms of posterior cerebral artery occlusions: A report of 2 cases. *Annals of Neurology* 2: 432–436.

Calkins, MW. (1893). A statistical study of pseudo-chromesthesia and of mentalforms. *American Journal of Psychology* 5: 439–464.

Calkins, MW. (1895). Synaesthesia. *American Journal of Psychology* 7: 90–107.

Callender, C. (2000). Shedding light on time. *Philosophy of Science* 67 (Proceedings): S587–S599.

Calvert, GA, et al. (1997). Activation of auditory cortex during silent lipreading. *Science* 276: 593–596.

Campen, C van. (2007). *The Hidden Sense: Synesthesia in Art and Science*. Cambridge: MIT Press.

Choo, CW. (1998). *Information Management for the Intelligent Organization.* Medford, NJ: Information Today.

Churchland, P. (1979). *Scientific Realism and the Plasticity of Mind.* Cambridge: Cambridge University Press.

Clancy, SA. (2005). *Abducted: How People Come to Believe They Were Kidnapped by Aliens.* Cambridge: Harvard University Press.

Cohen, L, et al. (2000). The visual word form area: Spatial and temporal characterization of an initial stage of reading in normal subjects and posterior split-brain patients. *Brain* 123: 291–307.

Cohen, L, Dehaene, S. (2004). Specialization within the ventral stream: The case for the visual word form area. *NeuroImage* 22: 466–476.

Corballis, MC. (1997). The genetics and evolution of handedness. *Psychological Review* 105: 714–777.

Cytowic, RE, Wood, FB. (1982). Synesthesia. II: Psychophysical relations in the synesthesia of geometrically shaped taste and colored hearing. *Brain and Cognition* 1: 36–49. Cytowic, RE. (1989). *Synesthesia: A Union of the Senses. New York: Springer Verlag (1st ed)*

Cytowic, RE. (1993). *The Man Who Tasted Shapes.* New York: Putnam.

Cytowic, RE. (1996). *The Neurological Side of Neuropsychology.* Cambridge: MIT Press.

Cytowic, RE. (1997). *Synesthesia: Phenomenology and Neuropsychology, pp 1–42,* in S Baron-Cohen, JE Harrison (eds.), *Synaesthesia: Classic and Contemporary Readings.* Oxford: Blackwell.

Cytowic, RE. (2002). *Synesthesia: A Union of the Senses,* 2nd ed. Cambridge: MIT Press.

Cytowic, RE. (2002). "Wahrnehnumgs–Synästhesie," pp 7–24, in H Adler & U Zeuch, eds, *Synästhesia: Interferenz–Transfer—Synthese der Sinne.* Würzburg: Königshausen & Neumann.

Cytowic, RE. (2003). The clinician's paradox: Believing those you must not trust. *Journal of Consciousness Studies* 10(9–10): 157–166.

Cytowic, RE. (2006). *My Auto-Neurography: A Memoir of Intellect, Emotion, and Detachment.* In preparation.

D'Andrade, R, Egan, M. (1975). The colors of emotion. *American Ethnologist* 1: 49–63.

Daily, A, Martindale, C, Borkum J. (1997). Creativity: Synesthesia and physiognomic perception. *Creativity Research Journal* 10: 1–8.

Dann, K. (1998). *Bright Colors, Falsely Seen: Synesthesia and the Search for Transcendental Knowledge.* New Haven: Yale University Press.

Day, S. (1996). Synesthesia and synesthetic metaphors. *PSYCHE: An Interdisciplinary Journal of Research on Consciousness* 2(32). http://psyche.cs.monash.edu.au/v2/psyche-2-32-day.html.

Day, S. (2005). Some demographic and socio-cultural aspects of synesthesia, pp. 11–33 in LC Robertson, N Sagiv (eds.), *Synesthesia: Perspectives from Cognitive Neuroscience*. New York: Oxford University Press.

Day, S. (2007, 3 December). *Types of Synesthesia.* http://home.comcast.net/~sean .day/html/types.htm.

de Gelder, B, Vroomen, J, Van der Heide, L. (1991). Face recognition and lip-reading in autism. *European Journal of Cognitive Psychology* 3: 69–86.

de Gelder, B, Bocker, KB, Tuomainen, J, Hensen, M, Vroomen, J. (1999). The combined perception of emotion from voice and face: Early interaction revealed by human electric brain responses. *Neuroscience Letters* 260: 133–136.

Dehaene, S, Bossini, S, Giraux, P. (1993). The mental representation of parity and numerical magnitude. *Journal of Experimental Psychology: General* 122: 371–396.

Dehaene, S, Dehaene-Lambertz, G, Cohen, L. (1998). Abstract representations of numbers in the animal and human brain. *Trends in Neuroscience* 21: 355–361.

Dehaene, S. (2001). Précis of the number sense. *Mind & Language* 16: 16–36.

Dehaene, S, Molko, L, Cohen, L, Wilson, A. (2004). Arithmetic and the brain. *Current Opinion in Neurobiology* 14: 218–224.

Devereaux, G. (1966). An unusual audio–motor synesthesia in an adolescent. *Psychiatric Quarterly* 40(3): 459–471.

Dixon, MJ, Smilek, D, Cudahy, C, Merikle, PM. (2000). Five plus two equals yellow: Mental arithmetic in people with synaesthesia is not coloured by visual experience. *Nature* 406(6794): 365.

Dixon, MJ, Smilek, D, Merikle, PM. (2004). Not all synaesthetes are created equal: Projector vs. associator synaesthetes. *Cognitive, Affective and Behavioral Neuroscience* 4: 335–343.

Domino, G. (1989). Synesthesia and creativity in fine art students: An empirical look. *Creativity Research Journal* 2(1–2): 17–29.

Doty, RL, Shaman, P, Dann, M. (1984). Development of the University of Pennsylvania Smell Identification Test: A standardized microencapsulated test of olfactory function. *Physiology & Behavior* (Monograph) 32: 489–502.

Downey, JE. (1911). A case of colored gustation. *American Journal of Psychology* 22: 528–539.

Dravnieks, A. (1985). *Atlas of Odor Character Profiles*. ASTM Data Series DS61. Philadelphia: AASTM.

Duchting, HJ. (1997). *Painting Music*. New York: Prestel, pp. 17, 65.

Dudycha, GJ, Dudycha, MM. (1935). A case of synesthesia: Visual-pain and visual-audition. *Journal of Abnormal and Social Psychology* 30: 57–69.

Duffy, PL. (2001). *Blue Cats and Chartreuse Kittens: How Synesthetes Color Their Worlds*. New York: Henry Holt.

Eagleman, DM. (2001). Visual illusions and neurobiology. *Nature Reviews Neuroscience* 2(12): 920–926.

Eagleman, DM. (2005). Obituary: Francis H. C. Crick. (1916–2004). *Vision Research* 45: 391–393.

Eagleman, DM, Kagan, AD, Nelson, SN, Sagaram, D, Sarma, AK. (2006). A standardized test battery for the study of synesthesia. *Journal of Neuroscience Methods* 159: 139–145.

Eagleman, DM, Nelson, S, Sarma, SK. (2007). The neuroscience, behavior, and genetics of synesthesia. Presentation, Society for Neuroscience.

Eagleman, DM. (In press). *Plasticity: How the Brain Rewires Itself on the Fly*. New York: Oxford University Press.

Eagleman, DM. (In press). *Ten Unsolved Questions of Neuroscience*. New York: Oxford University Press.

Edelman, G. (1992). *Bright Air, Brilliant Fire*. New York: Basic Books.

Edquist, J, Rich, AN, Brinkman, C, Mattingley, JB. (2006). Do synaesthetic colours act as unique features in visual search? *Cortex* 42: 222–231.

Emrich, HM, Schneider, U, Zeidler, M. (2000). *Welche Farbe hat der Montag?* Stuttgart: Hirzel Verlag.

English, H. (1923). Colored hearing. *Science* 57: 444.

Falchier, A, Clavagnier, S, Barone, P, Kennedy, H. (2002). Anatomical evidence of multimodal integration in primate striate cortex. *Journal of Neuroscience* 22: 5749–5759.

Feynman, RP. (1988). *What Do You Care What Other People Think?* New York: HarperCollins, p. 59.

Fias, W, Brysbaert, M, Geypens, F, D'ydewalle, G. (1996). The importance of magnitude information in numerical processing: Evidence from the SNARC effect. *Mathematical Cognition* 2: 95–110.

Fischinger, O. (1949). "True Creation," Knokke-le-Zoute Film Festival notes, reprinted in W Moritz. (2004). *Optical Poetry: The Life and Work of Oskar Fischinger*. London: John Libbey, p. 192.

Flechsig, P. (1901). Developmental (myelogenetic) localisation of the cerebral cortex in the human subject. *Lancet* 2, pp. 1027–1029.

Flournoy, T. (1893). *Des Phénomènes de Synopsie*, On the Phenomena of Synopsia. Geneva: Charles Eggimann.

Frith, U. (1989). *Autism: Explaining the Enigma*. Oxford: Blackwell.

Galton, F. (1880a). Visualized numerals. *Nature* 21: 252–256.

Galton, F. (1880b). Visualized numerals. *Nature* 22: 494–495.

Galton, F. (1883). *Enquiries into Human Faculty and Its Development*. London: Macmillian and Co.

Gazzaniga, MS, Eliassen, JC, Nisenson, L, Wessinger, CM, Baynes, KB. (1996). Collaboration between the hemispheres of a callosotomy patient: Emerging right hemisphere speech and the left brain interpreter. *Brain* 119: 1255–1262.

Gebhard, JW, Mowbray, GH. (1959). On discriminating the rate of visual flicker and auditory flutter. *American Journal of Psychology* 72: 521–528.

Gepner, B, et al. (1995). Postural effects of motion vision in young autistic children. *NeuroReport* 6: 1211–1214.

Gerland, G. (1997). *A Real Person: Life on the Outside* (J Tate, trans.). London: Souvenir Press.

Geschwind, N. (1964). The development of the brain and the evolution of language, pp. 155–169 in CJJM Stuart (ed.), *Monograph Series on Language and Linguistics, No. 17*. Washington, DC: Georgetown University.

Geschwind, N. (1965). Disconnection syndromes in animals and man, Part I (p. 275). *Brain* 88: 237–294.

Gloor, P. (1972). Temporal lobe epilepsy: Its possible contribution to the understanding of the significance of the amygdala and its interaction with neocortical–temporal mechanisms, pp. 423–457 in BE Eleftheriou (ed.), *The Neurobiology of the Amygdala*. New York: Plenum Press.

Gloor, P, Olivier, A, Quesney, LF, et al. (1982). The role of the limbic system in experiential phenomena of temporal lobe epilepsy. *Annals of Neurology* 12: 129–144.

Gloor, P. (1986). Role of the human limbic system in perception, memory, and affect: Lesions from temporal lobe epilepsy, pp. 159–169 in BK Doane, KE Livingston (eds.),

The Limbic System: Functional Organization and Clinical Disorders. New York: Raven Press.

Goléa, A. (1960). *Rencontres avec Olivier Messiaen*. Paris: Julliard.

Gray, JA, Williams, SCR, Nunn, J, et al. (1997). "Possible implications of synaesthesia for the hard question of consciousness," written at Malden, MA, in S Baron-Cohen, JE Harrison (eds.), *Synaesthesia: Classic and Contemporary Readings*. Oxford: Blackwell, pp. 173–181.

Gray, JA, Chopping, S, Nunn, J, et al. (2002). Implication of synaesthesia for functionalism. *Journal of Consciousness Studies* 9: 5–31.

Grossenbacher, PG, Lovelace, CT. (2001). Mechanisms of synesthesia: Cognitive and physiological constraints. *Trends in Cognitive Sciences* 5(1): 36–41.

Grossenbacher, PG. (1997). Perception and sensory information in synesthetic experience, in S Baron-Cohen, J Harrison (eds.), *Synaesthesia: Classic and Contemporary Readings*. Oxford: Blackwell, pp. 148–172.

Haber, RN, Haber, RB. (1964). Eidetic imagery. I: Frequency. *Perceptual and Motor Skills* 19: 131–138.

Haber, RN. (1969). Eidetic images. *Scientific American* 220: 36–44.

Haber, RN. (1979). Twenty years of haunting eidetic imagery: Where's the ghost? *Behavioral and Brain Sciences* 2: 583–629.

Hall, GS. (1883). The contents of children's minds. *Princeton Review* 11(4th serial): 249–272.

Happé, FGE. (1996). Studying weak central coherence at low levels: Children with autism do not succumb to visual illusions: A research note. *Journal of Child Psychology and Psychiatry* 37: 873–877.

Happé, F. (1999). Autism: Cognitive deficit or cognitive style? *Trends in Cognitive Sciences* 3(6): 216–222.

Harrison, JE, Baron-Cohen, S. (1997). pp. 109–122 in S Baron-Cohen, JE Harrison (eds.), *Synaesthesia: Classic and Contemporary Readings*. Oxford: Blackwell.

Harrison, JE. (2001). *Synaesthesia: The Strangest Thing*. New York: Oxford University Press.

Hausser-Hauw, C, Bancaud, J. (1987). Gustatory hallucinations in epileptic seizures: Electrophysiological, clinical and anatomical correlates. *Brain* 110(Pt. 2): 339–359.

Helson, H. (1933). A child's spontaneous reports of imagery. *American Journal of Psychology* 45: 360–361.

Henik, A, Tzelgov, J. (1982). Is three greater than five: The relation between physical and semantic size in comparison tasks. *Memory and Cognition* 10: 389–395.

Herman, L, Richards, D, Wolz, J. (1984). Comprehension of sentences by bottle nosed dolphins. *Cognition* 16: 129–219.

Heron, W. (1957). The pathology of boredom. *Scientific American* 196: 52–56.

Holden, ES. (1891). Colour-associations with numerals etc. *Nature* 44: 223–224.

Horowitz, MJ. (1975). Hallucinations: An information processing approach, pp. 163–196, in RK Siegel, LJ West (eds.), *Hallucinations: Behavior, Experience and Theory*. New York: Wiley.

Howard, RJ, et al. (1998). The functional anatomy of imagining and perceiving colour. *NeuroReport* 9: 1019–1023.

Hubbard, EM, Arman, AC, Ramachandran, VS, Boynton, GM. (2005). Individual differences among grapheme–color synesthetes: Brain–behavior correlations. *Neuron* 45(6): 975–985.

Hubbard, EM, Piazza, M, Pinel, P, Dehaene, S. (2005). Interactions between number and space in parietal cortex. *Nature Reviews Neuroscience* 6(6):435–448.

Hubbard, EM, Ramachandran, VS. (2005). Neurocognitive mechanisms of synesthesia. *Neuron* 48: 509–520.

Hubbard, EM, Manohar, S, Ramachandran, VS. (2006). Contrast affects the strength of synesthetic colors. *Cortex* 42: 184–194.

Hubbard, TL. (1996). Synesthesia-like mappings of lightness, pitch, and melodic interval. *American Journal of Psychology* 109: 219–238.

Hunt, HT. (2005). Synaesthesia, metaphor, and consciousness. *Journal of Consciousness Studies* 12(12): 26–45.

Innocenti, GM. (1986). General organization of callosal connections in the cerebral cortex, in EG Jones, A Peters (eds.), *Cerebral Cortex*, vol. 5. New York: Plenum, pp. 291–353.

Jacobs, L, Karpik, A, Bozian, D, Gøthgen, S. (1981). Auditory–visual synesthesia: Sound-induced photisms. *Archives of Neurology* 38(4): 211–216.

Jacome, DE, Gumnit, RJ. (1979). Audioalgesic and audiovisuoalgesic synesthesias: Epileptic manifestation. *Neurology* 29: 1050–1053.

Jaensch, ER. (1930). *Eidetic Imagery and Typological Methods of Investigation*, 2nd ed. (O Oeser, trans.). New York: Harcourt Brace.

James, W. (1890). *The Principles of Psychology*. New York: Dover.

Jarrold, C, Russell, J. (1997). Counting abilities in autism: Possible implications for central coherence theory. *Journal of Autism and Developmental Disorders* 27: 25–37.

Jiyu-Kennett, PTNH. (1990). The scripture of great wisdom, in *The Liturgy of the Order of Buddhist Contemplatives for the Laity*. Mt. Shasta, CA: Shasta Abbey Press, pp. 73–74.

Johnson, DB. (1985). *Worlds in Regression: Some Novels of Vladimir Nabokov*. Ann Arbor: Ardis.

Jordan, DS. (1917). The colors of letters. *Science* 46(1187): 311–312.

Kadosh, RC, Sagiv, N, Linden, EJ, et al. (2005). When blue is larger than red: Colors influence numerical cognition in synesthesia. *Journal of Cognitive Neuroscience* 17(11): 1766–1773.

Kennedy, H, Batardiere, A, Dehay, C, Barone, P. (1997). Synesthesia: Implications for developmental neurobiology, pp. 243–258 in S Baron-Cohen, JE Harrison (eds.), *Synaesthesia: Classic and Contemporary Readings*. Oxford: Blackwell.

Kloos, G. (1931). Synästhesien bei psychisch Abnormen. Eine Studie über das Wesen der Synästhesie und der synästhetischen Anlage. *Archiv für Psychiatrie und Nervenkrankheiten* 94: 418–469.

Klüver, H. (1966). *Mescal and Mechanisms of Hallucinations*. Chicago: University of Chicago Press, p. 22.

Köhler, W. (1929/1947). *Gestalt Psychology*, 2nd ed. New York: Liveright.

Komisaruk, BR, Whipple, B. (2005). Functional MRI of the brain during orgasm in women. *Annual Review of Sex Research* 16: 62–86.

Lakoff, G, Johnson, MH. (1980). *Metaphors We Live By*. Chicago: University of Chicago Press.

Lenay, C, Gapenne, O, Hanneton, S, Marque, C, Genouel, C. (2003). Sensory substitution: Limits and perspectives, in *Touching for Knowing: Cognitive Psychology of Haptic Manual Perception*. Amsterdam/Philadelphia: John Benjamins, pp. 275–292.

Lepore, F. (1990). Spontaneous visual phenomena with visual loss: 104 patients with lesions of retinal and neural afferent pathways. *Neurology* 40: 444–447.

Lewkowicz, D, Turkewitz, G. (1980). Cross-modal equivalence in infancy: Auditory–visual intensity matching. *Developmental Psychology* 16: 597–607.

Lickliter, R, Bahrick, LE. (2004). Perceptual development and the origins of multisensory responsiveness, pp. 643–654 in G Calvert, C Spence, BE Stein (eds.), *Handbook of Multisensory Processes*. Cambridge: MIT Press.

Liu, H, Hockenberry, M, Selker, T. (2005). http://web.media.mit.edu/~hugo/publications/papers/SIGGRAPH2005-SynaestheticRecipes.pdf.

Loe, PR, Benevento, LA. (1969). Auditory–visual interaction in single units in the orbito–insular cortex of the cat. *Electroencephalography and Clinical Neurophysiology* 26: 395–398.

Luria, AR. (1968). *The Mind of a Mnemonist*. New York: Basic Books.

Macaluso, E, Frith, CD, Driver, J. (2000). Modulation of human visual cortex by crossmodal spatial attention. *Science* 289: 1206–1208.

MacLeod, CM. (1991). Half a century of research on the Stroop effect: An integrative review. *Psychological Bulletin* 109: 163–203.

Maddock, RJ, Buonocore, MH. (1997). Activation of the left posterior cingulate gyrus by the auditory presentation of threat related words: An fMRI study. *Psychiatry Research* 75: 1–14.

Maddock, RJ. (1999). The retrosplenial cortex and emotion: New insights from functional neuroimaging of the human brain. *Trends in Neuroscience* 22: 310–316.

Mahling, F. (1926). Das Problem der "Audition colorée": Eine historische-kritische Untersuchung. *Archiv für Gesamte Psychologie* 57: 165–301.

Marks, LE. (1974). On associations of light and sound: The mediation of brightness, pitch, and loudness. *American Journal of Psychology* 87: 173–188.

Marks, LE. (1975). On colored–hearing synesthesia: Cross-modal translations of sensory dimensions. *Psychological Bulletin* 82(3): 303–331.

Marks, LE. (1978). *The Unity of the Senses: Interrelations among the Modalities*. New York: Academic Press.

Marks, LE, Hammeal, RJ, Bornstein, MH. (1987). Perceiving similarity and comprehending metaphor. *Monographs of the Society for Research in Child Development* 52: 1–93.

Marks, LE, Bornstein, MH. (1987). Sensory similarities: Classes, characteristics, and cognitive consequences, pp. 49–65, in RE Haskell (ed.), *Cognition and Symbolic Structures: The Psychology of Metaphoric Transformation*. Norwood, NJ: Ablex.

Marks, LE. (1989). On cross-modal similarity: The perceptual structure of pitch, loudness, and brightness. *Journal of Experimental Psychology: Human Perception and Performance* 15: 586–602.

Marks, LE. (2004). Cross-modal interactions in speeded classification, pp. 85–106 in G Calvert, C Spence, BE Stein (eds.), *Handbook of Multisensory Processes*. Cambridge: MIT Press.

Mass, W. (2003). *A Mango-Shaped Space*. Little, Brown.

Mattingley, JB, Ward, J (eds.). (2006). *Cognitive Neuroscience Perspectives on Synaesthesia* (Special Issue). *Cortex* 42: 129–320.

Maurer, D. (1997). Neonatal synesthesia: Implications for the processing of speech and faces, pp. 224–242 in S Baron-Cohen, JE Harrison (eds.), *Synaesthesia: Classic and Contemporary Readings*. Oxford: Blackwell.

Maurer, D, Pathman, T, Mondloch, CJ. (2006). The shape of boubas: Sound–shape correspondences in toddlers and adults. *Developmental Science* 9(3): 316–322.

McGurk, G, MacDonald, J. (1976). Hearing lips and seeing voices. *Nature* 264: 746–748.

McKellar, P. (1957). *Imagination and Thinking*. New York: Basic Books.

McKelvie, SJ, Rohrberg, MM. (1978). Individual differences in reported visual imagery and cognitive performance. *Perceptual and Motor Skills* 46(2): 451–458.

McTaggart, JME. (1908). The Unreality of Time. *Mind* 18: 457–484.

Meltzoff, A, Moore, M. (1992). Early imitation within a functional framework. *Infant Behavior and Development* 15: 479–505.

Meredith, MA, Nemitz, JW, Stein, BE. (1987). Determinants of multisensory integration in superior colliculus neurons. I: Temporal factors. *Journal of Neuroscience* 7: 3215–3229.

Messiaen, O. (1944). *The Technique of My Musical Langauge*, vol. 1 (J Satterfield, trans.). Paris: Alphonse Leduc, p. 51.

Messiaen, O. (1977). *Aux canyons des etoiles*. Liner notes, Erato STU70974/975 (recording). Paris: Alphonse Leduc.

Mesulam, M-M. (2000). *Principles of Behavioral and Cognitive Neurology*, 2nd ed., pp. 1–120. New York: Oxford University Press.

Miller, TC, Crosby, TW. (1979). Musical hallucinations in a deaf elderly patient. *Annals of Neurology* 5: 301–302.

Mills, CB, Boteler, EH, Oliver, GK. (1999). Digit synaesthesia: A case study using a Stroop-type test. *Cognitive Neuropsychology* 16: 181–191.

Morgan, GA, Goodson, FE, Jones, T. (1975). Age differences in the associations between felt temperatures and color choices. *American Journal of Psychology* 88(1): 125–130.

Moritz, W. (2004). *Optical Poetry: The Life and Work of Oskar Fischinger*. London: John Libbey, p. 84.

Myers, CS. (1914–1915). Two cases of synaesthesia. *British Journal of Psychology* 7: 112–117.

Myles, KM, Dixon, MJ, et al. (2003). Seeing double: The role of meaning in alpha-numeric–colour synaesthesia. *Brain and Cognition* 53(2): 342–345.

Nabokov, V. (1949). Portrait of my mother. *New Yorker* April 9, pp. 33–37.

Nabokov, V. (1962). *Pale Fire*. New York: Putnam, p. 34.

Nabokov, V. (1966). *Speak, Memory: An Autobiography Revisited*. New York: Dover. (First published in 1951 as *Conclusive Evidence*.)

Nabokov, V. (1969). *Ada or Ardor: A Family Chronicle*. New York: McGraw-Hill, p. 584.

Nagel, T. (1986). *The View from Nowhere*. New York: Oxford University Press.

Neville, HJ. (1995). Developmental specificity in neurocognitive development in humans, in M Gazzaniga (ed.), *The Cognitive Neurosciences*. Cambridge: MIT Press.

Nikolić, D, Lichti, P, Singer, W. (2007). Color-opponency in synaesthetic experiences. *Psychological Science* 18(6): 481–486.

Nunn, JA, Gregory, LJ, Brammer, M, Williams, SCR, Parslow, DM, Morgan, MJ, et al. (2002). Functional magnetic resonance imaging of synesthesia: Activation of V4/V8 by spoken words. *Nature Neuroscience* 5: 371–375.

Osgood, CE. (1952). The nature and measurement of meaning. *Psychological Bulletin* 49: 197–237.

Osterbauer, RA, Matthews, PM, Jenkinson, M, Beckmann, CF, Hansen, PC, Calvert, GA. (2005). Color of scents: Chromatic stimuli modulate odor responses in the human brain. *Journal of Neurophysiology* 93(6): 3434–3441.

Palmeri, TJ, Blake, R, Marois, R, et al. (2002). The perceptual reality of synesthetic colors. *Proceedings of the National Academy of Sciences, USA* 99: 4127–4131.

Pascual-Leone, A, Amedi, A, Fregni, F, Merabet, LB. (2005). The plastic human brain cortex. *Annual Review of Neuroscience* 28: 377–401.

Patterson, KE, Morton, J. (1985). From orthography to phonology: An attempt at an old interpretation, pp. 335–359 in KE Patterson, JC Marshall, M Coltheart (eds.), *Surface Dyslexia*. Hillsdale, NJ: Lawrence Erlbaum.

Paulesu, E, Harrison, J, Baron-Cohen, S, et al. (1995). The physiology of coulored hearing: A PET activation study of coloured-word synaesthesia. *Brain* 118: 661–676.

Paulsen, HG, Laeng, B. (2006). Pupillometry of grapheme–color synaesthesia. *Cortex* 42: 290–294.

Peacock, K. (1988). Instruments to perform color-music: Two centuries of techno-logical instrumentation. *Leonardo* 21: 397–406.

Pepperberg, I. (1999). *The Alex Studies: Cognitive and Communicative Abilities of Grey Parrots*. Cambridge: Harvard University Press.

Persinger, MA. (1983). Religious and mystical experiences as artifacts of temporal lobe function: A general hypothesis. *Perceptual and Motor Skills* 57: 1255–1262.

Petersen, SE, Fox, PT, Snyder, AZ, Raichle, ME. (1990). Activation of extrastriate and frontal cortical areas by visual words and word-like stimuli. *Science* 249: 1041–1044.

Piazza, M, Pinel, P, Dehaene, S. (2006). Objective correlates of a peculiar subjective experience: A single-case study of number–form synaesthesia. *Cognitive Neuropsychology* 23(8): 1081–1082.

Pierce, AH. (1907). Gustatory audition: A hitherto undescribed variety of synaesthesia. *American Journal of Psychology* 18: 341–352.

Plodowski, A, Swainson, R, Jackson, GM, Rorden, C, Jackson, SR. (2003). Mental representation of number in different numerical forms. *Current Biology* 13(23): 2045–2050.

Plummer, HC. (1915). Color music—A new art created with the aid of science: The color organ used in Scriabin's symphony "Prometheus." *Scientific American* (April 10).

Popper, KR, Eccles, JC. (1977). *The Self and Its Brain*. New York: Springer Verlag, p. 469.

Pritchard, TC, Macaluso, DA, Eslinger, PJ. (1999). Taste perception in patients with insular cortex lesions. *Behavioral Neuroscience* 113: 663–671.

Profitta, J, Bidder, H. (1988). Perfect pitch. *Journal of Musical Genetics* 29(4): 763–771.

Purpura, DP. (1956a). Electrophysiological analysis of psychotogenic drug action. I: Effect of lysergic acid diethylamide (LSD) on specific afferent systems in the cat. *Archives of Neurology and Psychiatry* 75: 122–131.

Purpura, DP. (1956b). Electrophysiological analysis of psychotogenic drug action. II: General nature of lysergic acid diethylamide (LSD) action on central synapses. *Archives of Neurology and Psychiatry* 75: 132–143.

Purpura, DP. (1957). Experimental analysis of the inhibitory action of LSD on cortical dendritic activity. *Annals of the New York Academy of Sciences* 66: 515–536.

Ramachandran, VS, Hirstein, WS, Armel, KC, et al. (1997). The neural basis of religious experience. *Society for Neuroscience Abstracts* 23: 1316.

Ramachandran, VS, Hubbard, EM. (2001). Psychophysical investigations into the neural basis of synaesthesia. *Proceedings of the Royal Society of London B* 268: 979–983.

Ramachandran, VS, Hubbard, EM. (2001). Synaesthesia—A window into perception, thought, and language. *Journal of Consciousness Studies* 8(12): 3–34.

Ramachandran, VS, Hubbard, EM. (2003). The phenomenology of synaesthesia. *Journal of Consciousness Studies* 10: 49–57.

Ramachandran, VS. (2004). *A Brief Tour of Human Consciousness.* New York: Pi Press, p. 74.

Ramachandran, VS, Hubbard, EM, Butcher, PA. (2004). Synesthesia, cross-activation, and the foundations of neuroepistemology, pp. 867–883 in G Calvert, C Spence, BE Stein (eds.), *Handbook of Multisensory Processes.* Cambridge: MIT Press.

Raven, JC. (1956). *Guide to Using the Raven's Progressive Matrices.* London: HE Lewis.

Renkel, M. (1957). Pharmacodynamics of LSD and mescaline. *Journal of Nervous and Mental Diseases* 125: 424–427.

Rich, AN, Mattingley, JB. (2002). Anomalous perception in synaesthesia: A cognitive neuroscience perspective. *Nature Reviews Neuroscience* 3(1): 43–52.

Rich, AN, Bradshaw, JL, Mattingley, JB. (2005). A systematic large-scale study of synaesthesia: Implications for the role of early experience in lexical–color associations. *Cognition* 98: 53–84.

Riggs, L, Karwoski, T. (1934). Synaesthesia. *British Journal of Psychology* 25: 29–41.

Rizzo, MR, Esslinger, PJ. (1989). Colored hearing synesthesia: Investigation of neural factors in a single subject. *Neurology* 39: 781–784.

Robertson, L, Sagiv, N (eds.). (2005). *Synesthesia: Perspectives from Cognitive Neuroscience.* Oxford: Oxford University Press.

Rolls, ET, Bayliss, LL. (1994). Gustatory, olfactory and visual convergence within the primate orbitofrontal cortex. *Journal of Neuroscience* 14: 5437–5452.

Root-Bernstein, R, Root-Bernstein, M. (1999). *Sparks of Genius: The Thirteen Thinking Tools of the World's Most Creative People.* Boston: Houghton Mifflin.

Rouw, R, Scholte, HS. (2007). Increased structural connectivity in grapheme–color synesthesia. *Nature Neuroscience* 10: 792–797.

Royet, JP, Koenig, O, Gregoire, MC, et al. (1999). Functional anatomy of perceptual and semantic processing for odors. *Journal of Cognitive Science* 11(1): 94–109.

Royet, JP, Hudry J, Zald, JH, et al. (2001). Functional neuroanatomy of different olfactory judgments. *Neuroimage* 13: 506–519.

Russell, B. (1915). On the Experience of Time. *The Monist* 25: 212–233.

Sadato, N, Pascual-Leone, A, Grafman, J, Ibanez, V, Deiber, MP, et al. (1996). Activation of the primary visual cortex by Braille reading in blind subjects. *Nature* 380: 526–528.

Sadato, N, Pascual-Leone, A, Grafman, J, Deiber, MP, Ibanez, V, Hallett, M. (1998). Neural networks for Braille reading by the blind. *Brain* 121(Pt. 7): 1213–1229.

Sagiv, N, Heer, J, Robertson, LC. (2006a). Does binding of synesthetic color to the evoking grapheme require attention? *Cortex* 42: 232–242.

Sagiv, N, Simner, J, Collins, J, Butterworth, B, Ward, J. (2006b). What is the relationship between synaesthesia and visuo–spatial number forms? *Cognition* 101(1): 114–128.

Sagiv, N, Olu-Lafe, O, Amin, M, Ward, J. (2006c). Grapheme personification: A profile. UK Synaesthesia Association 2nd Annual Meeting. April 22–23. London, UK.

Samuel, C. (1976). *Conversations with Olivier Messiaen* (F Aprahamian, trans.). London: Stainer and Bell.

Schlaug, G, Jancke, L, Huang, Y, Steinmetz, H. (1995). In vivo evidence of structural brain asymmetry in musicians. *Science* 267: 699–701.

Schomer, DL, O'Connor, M, Spiers, P, et al. (2000). Temporolimbic epilepsy and behavior, pp. 377–388 in M-M Mesulam (ed.), *Principles of Behavioral and Cognitive Neurology*. New York: Oxford University Press.

Schultze, E. (1912). Krankhafter Wandertrieb, räumlich beschränkte Taubheit für bestimmte Töne und "tertiare" Empfindungen bei einem Psychopathen. *Zeitschrift für die gesamte Neurologie und Psychiatrie* 10: 399.

Schwartz, J, Robert-Ribes, J, Escudier, JP. (1998). Ten years after Summerfield: A taxonomy of models for audiovisual fusion in speech perception, p. 319 in R Campbell, B Dodd, DK Burnham (eds.), *Hearing by Eye*. East Sussex: Hove.

Scriabin, A. (1911). *Prometheus, The Poem of Fire Opus 60.*

Seitz, JA. (2005). The neural, evolutionary, developmental, and bodily basis of metaphor. *New Ideas in Psychology* 23: 74–95.

Seron, X, Peseenti, M, Noel, M-P, et al. (1992). Images of numbers or "when 98 is upper left and 6 is sky blue." *Cognition* 44: 159–196.

Shah, NJ, Marshall, JC, Safiris, O, et al. (2001). The neural correlates of person familiarity: A functional MRI imaging study with clinical applications. *Brain* 124: 804–815.

Shams, L, Kamitani, Y, Shimojo, S. (2000). Illusions: What you see is what you hear. *Nature* 408(6814): 788.

Shanon, B. (1982). Color associates to semantic linear orders. *Psychological Research* 44: 75–83.

Shipley, T. (1964). Auditory flutter-driving of visual flicker. *Science* 145: 1328–1330.

Shuler, MG, Bear, MF. (2006). Reward timing in the primary visual cortex. *Science* 311(5767): 1606–1609.

Siegel, RK, Jarvik, ME. (1975). Drug-induced hallucinations in animals and man, pp. 81–162 in RK Siegel, LJ West (eds.), *Hallucinations: Behavior, Experience, and Theory*. New York: Wiley.

Siegel, RK. (1977). Hallucinations. *Scientific American* 237(4): 132–140.

Simner, J, Ward, J, Lanz, M, Jansari, A, Noonan, K, Glover, L, Oakley, D. (2005). Non-random associations of graphemes to colors in synaesthetic and normal populations. *Cognitive Neuropsychology* 22(8): 1069–1085.

Simner, J, Mulvenna, C, Sagiv, N, Tsakanikos, E, Witherby, SA, Fraser, C, Scott, K, Ward, J. (2006). Synaesthesia: The prevalence of atypical cross-modal experiences. *Perception* 35(8): 1024–1033.

Simner, J, Glover, L, Mowat, A. (2006). Linguistic determinants of word colouring in grapheme–colour synaesthesia. *Cortex* 42: 281–289.

Simner, J, Hubbard, EM. (2006). Variants of synesthesia interact in cognitive tasks: Evidence for implicit associations and late connectivity in cross-talk theories. *Neuroscience* 143(3): 805–914.

Simner, J, Holenstein, E. (2007). Ordinal linguistic personification as a variant of synesthesia. *Journal of Cognitive Neuroscience* 19(4): 694–703.

Simpson, R, Quinn, M, Ausubel, D. (1956). Synesthesia in children: Associations of colors with pure tone frequencies. *Journal of Genetic Psychology* 89: 95–103.

Smilek, D, Dixon, MJ, Cudahy, C, Merikle, PM. (2001). Synaesthetic photisms influence visual perception. *Journal of Cognitive Neuroscience* 13: 930–936.

Smilek, D, Dixon, MJ, Cudahy C, Merikle, PM. (2002). Concept driven color experiences in digit–color synesthesia. *Brain and Cognition* 48(2–3): 570–573.

Smilek, D, Moffatt, BA, Pasternak, J, White, BN, Dixon, MJ, Merikle, PM. (2002). Synaesthesia: A case study of discordant monozygotic twins. *Neurocase* 8: 338–342.

Smilek, D, Dixon, MJ, Cudahy, C, Merikle, PM. (2002). Synesthetic color experiences influence memory. *Psychological Science* 13(6): 548–552.

Smilek, D, Callejas, A, Merikle, P, Dixon, M. (2006). Ovals of time: Space–time synesthesia. *Consciousness and Cognition* 16(2): 507–519.

Smilek, D, Malcolmson, KA, Carriere, JS, Eller, M, Kwan, D, Reynolds, M. (2007). When "3" is a jerk and "E" is a king: Personifying inanimate objects in synesthesia. *Journal of Cognitive Neuroscience* 19(6): 981–992.

Sperling, JM, Prvulovic, D, Linden, DEJ, et al. (2006). Neuronal correlates of colour–graphemic syaesthesia: A fMRI study. *Cortex* 42: 295–303.

Starr, F. (1893). Note on colored hearing. *American Journal of Psychology* 51: 416–418.

Steen, CJ. (2001). Visions shared: A firsthand look into synesthesia and art. *Leonardo* 34: 203–208.

Stein, BE, Meredith, MA. (1993). *The Merging of the Senses*. Cambridge: MIT Press.

Stevenson, RJ, Boakes, RA. (2004). Sweet and sour smells: Learned synesthesia between the senses of taste and smell, pp. 69–84 in G Calvert, C Spence, BE Stein (eds.), *Handbook of Multisensory Processes*. Cambridge: MIT Press.

Stromeyer, CF, Psotka, J. (1970). The detailed texture of eidetic images. *Nature* 225: 346–349.

Stroop, JR. (1935). Studies of interference in serial verbal reactions. *Journal of Experimental Psychology* 28: 643–662.

Suarez de Mendoza, F. (1890). *L'audition colorée*. Paris: Octave Donin.

Sullivan, JWN. (1914). An organ on which color compositions are played: The new art of color music and its mechanism. *Scientific American* (February 21).

Swinkels, WAM, Kuyk, J, van Dyck, J, Spinhoven, PH. (2005). Psychiatric comorbidity in epilepsy. *Epilepsy and Behavior* 7: 37–50.

Tellegen, A, Atkinson, G. (1978). Openness to absorbing and self altering experiences ("absorption"): A trait related to hypnotic susceptibility. *Journal of Abnormal Psychology* 83: 268–277.

Ternaux, JP. (2003). Synesthesia: A multimodal combination of senses. *Leonardo* 36: 321–322.

Teuber, HL. (1961). Sensory deprivation, sensory suppression and agnosia: Notes for neurologic theory. *Journal of Nervous and Mental Diseases* 132: 32–40.

Torrance, JP. (1966). *Thinking Creatively with Words*. Princeton: Personnel Press.

Toth, LJ, Assad, JA. (2002). Dynamic coding of behaviourally relevant stimuli in parietal cortex. *Nature* 415(6868): 165–168.

Tzourio-Mazoyer, N, De Schonen, S, Crivello, F, et al. (2002). Neural correlates of woman face processing by 2-month-old infants. *Neuroimage* 15: 454–461.

Kandkisky, W. (1912). Über das Geistige in der Kunst, trans. M. Sadleir, *On the Spiritual in Art* (1947). New York: Wittenborn, Schultz.

Van Orden, GC. (1987). A rows is a rose: Spelling, sound, and reading. *Memory and Cognition* 14: 371–386.

Van Wassenhove, V, Grant, KW, Poeppel, D. (2007). Temporal window of integration in auditory–visual speech perception. *Neuropsychologia* 45(3): 598–607.

Vike, J, Jabbari, B, Maitland, CG. (1984). Auditory–visual synesthesia: Report of a case with intact visual pathways. *Archives of Neurology* 41: 680–681.

Vroomen, J, de Gelder, B. (2004). Perceptual effects of cross-modal stimulation: Ventriloquism and the freezing phenomenon, pp. 141–150 in G Calvert, C Spence, BE Stein (eds.), *Handbook of Multisensory Processes*. Cambridge: MIT Press.

Vygotsky, L. (1965). *Thought and Language*. Cambridge: MIT Press.

Wallace, MT. (2004). The development of multisensory integration, pp. 625–642 in G Calvert, C Spence, BE Stein (eds.), *Handbook of Multisensory Processes*. Cambridge: MIT Press.

Walsh, R. (2005). Can synaesthesia be cultivated: Indications from surveys of meditators. *Journal of Consciousness Studies* 12(4–5): 5–17.

Ward, J, Collins, J, Auyeung, V. (2003). Word–taste synaesthesia is an automatic and perceptual phenomenon. *Journal of Cognitive Neuroscience* 15(suppl): 51.

Ward, J, Simner, J. (2003). Lexical–gustatory synaesthesia: Linguistic and conceptual factors. *Cognition* 89: 237–261.

Ward, J. (2004). Emotionally mediated synaesthesia. *Cognitive Neuropsychology* 21(7): 761–772.

Ward, J, Simner, J, Auyeung, V. (2005). A comparison of lexical–gustatory and grapheme–colour synaesthesia. *Cognitive Neuropsychology* 22(1): 28–41.

Ward, J, Simner, J. (2005). Is synaesthesia an X-linked trait with lethality in males? *Perception* 34: 611–623.

Ward, J, Tsakanikos, E, Bray, A. (2006). Synaesthesia for reading and playing musical notes. *Neurocase* 12(1): 27–34.

Ward, J, Huckstep, B, Tsakanikos, E. (2006). Sound–colour synaesthesia: To what extent does it use cross-modal mechanisms common to us all? *Cortex* 42: 264–280.

Weiss, PH, Shah, NJ, Toni, I, Zilles, K, Fink, GR. (2001). Associating colours with people: A case of chromatic-lexical synaesthesia. *Cortex* 37: 750–753.

Weiss, PH, Zilles, K, Fink, GR. (2005). When visual perception causes feeling: Enhanced cross-modal processing in grapheme–color synesthesia. *NeuroImage* 28: 859–868.

Welch, RB, DuttonHurt, LD, Warren, DH. (1986). Contributions of audition and vision to temporal rate perception. *Perception & Psychophysics* 39: 294–300.

Wellek, A. (1931). Zur Geschichte und Kritik der Synästhesie-Forschung. *Archiv für die gesamte Psychologie* 79: S.325–384.

Werner, H. (1940). *Comparative Psychology of Mental Development*. New York: Harper.

Wheeler, RH, Cutsforth, TD. (1921). The number forms of a blind subject. *American Journal of Psychology* 32: 21–25.

Whitchurch, AK. (1922). Synesthesia in a child of three and a half years. *American Journal of Psychology* 33: 302–303.

Wild, T, Kuiken, D, Schopflocher, D. (1995). The role of absorption in experiential involvement. *Journal of Personality and Social Psychology* 69: 569–579.

Witthoft, N, Winawer, J. (2006). Synesthetic colors determined by having colored refrigerator magnets in childhood. *Cortex* 42: 175–183.

Wynn, K. (1990). Children's understanding of counting. *Cognition* 36(2): 155–193.

Yakovlev, P. (1962). Morpholological criteria of growth and maturation of the nervous sysem in man. *Research Publications of the Association for Nervous and Mental Disorders* 39: 3–46.

Yakovlev, PI, Lecours, AR. (1967). The myelogenetic cycles of regional maturation of the brain, pp. 3–70 in A Minkowski (ed.), *Regional Development of the Brain in Early Life*. Oxford: Blackwell.

Zeki, S. (1993). *A Vision of the Brain*. Oxford: Blackwell Science.

Ziegler, MJ. (1930). Tone shapes: A novel type of synesthesia. *Journal of General Psychology* 3: 227–287.

Index

Note: An "f" after the page number denotes a figure, a "t" denotes a table.

"A" as commonly red, 77–80
Absorption, 172, 194–195, 215
Absolute pitch, 95–97
Acupuncture, 178, 180
Acoustic properties (speech/music), 39, 87, 95, 97, 144–145
Acquired synesthesia, 38f, 39, 215–224
Ada, 54, 175–176, 253
Adjective descriptors, 7, 129, 133
Affect, 54–56, 61, 160. *See also* Emotion
widened, 161
Afterimages, 53
Alcohol, effect on synesthesia, 103, 138, 139f, 146, 224
Alcohol withdrawal, effect of, 138–139
Alien color effect, 63, 140–145, 149
taste–reading, 146
Allochiria, 244
Allophony, 148
Alzheimer's, 153
Ambiguous stimuli, 74–76. *See also* Contextual effects
American Synesthesia Association, 227
Amphetamine, 138, 224
Amyl nitrite, 137–138, 224
Angular gyrus, 195–196, 212, 243–244
Anosmia, 96
Armel, Carrie, 204, 219

Aroma network, 128, 129f
Art, 163, 172–198, 174f
Attention, 48–49, 73–76, 107
affecting color intensity, 74
attentional load, 50
looking vs. seeing, 74
shift of, 34, 76, 76f
Audio-motor synesthesia, 40, 43, 99, 103
Auditory driving illusion, 203
Auras, 10, 151–157. *See also* Emotionally mediated synesthesia
in migraine, 57, 88
Autism, as opposite of synesthesia, 241–242
Automaticity, 47–50, 63, 104
Autoscopy, 160
Aux Canyons des Etoiles, 94

Bach, Johann Sebastian, 90, 96, 191
Baron-Cohen, Simon, 7, 77, 102, 122
Bashō, 189–190
Baudelaire, Charles, 188
Bauhaus, 188
Beach, Amy, 13, 93
Beethoven, Ludwig von, 14, 187
Behaviorism, 15–16, 235–236
Berlin and Kay order of color names, 78–79, 156, 189
Bernard, Jonathan, 94
Biases of theory, 17–18, 20

Bidirectional. *See* Synesthesia
Binding problem, 201–203
Blavatsky, Madame, 190
Blindfolded subjects, 204–205, 216
Blindness, 39, 45, 54, 96, 204, 218–219
Blindsight, 44–45
Body language, 132, 152
Boredom, 218
Bouba-kiki, 165–166, 195
Bowerman, Jane, 89, 89f, 173, 174f
Braille, 32, 45, 90, 204–206, 216
Brain development, 13, 35–36, 79, 196
Brain plasticity. *See* Plasticity
Brodmann, Korbinian, 230
Brumbaugh, R. S., 82
Bryce canyon, 94
Brydon, Bruce, 10, 151–152
Bush, Rita, 44, 44f, 244

Canaletto, 185
Change (drift) in color, 67–71, 68f,
    152, 237–238
Charlatans, 156
Chaos theory, 61
Chicken à la mode, 140
Chicken, pointed, 3
Circle of fifths, 190–191, 191f
Chords (music), 91, 94–95, 143, 173,
    184
Clairvoyance, sense of, 154, 160
Clavier a lumiéres, 190
Clocklike patterns, 17–18, 31f
Chronochromie, 173
Color conflict. *See* Alien color effect
Colored equations, 28
Colored graphemes, 30–34, 36, 68.
    See also Graphemes
    frequency of, 77
Colored hearing, 14, 26, 36, 39, 39f,
    87–95, 102, 136
Colorless graphemes, 66, 77, 238–239
    auditory influence, 68
Color names, fixed order of, 7–8, 79

Color, predominance of, 239
Colored music, conventional, 7, 190
Colored persons, 153–154
Colored phonemes, 99–101. *See also*
    Phonemes
Colored tastes and smells, 140–145
Color notation on musical scores, 174
Color vision, normal, 6
Color priming, 47–48, 48f, 164, 164f
Colors of light vs. pigment, 178
Color shades. *See* Shades
Colours de la citie celeste, 174
Compound sensation, 127
Computerized color matching, 65–66
Conceptual determinants, 70–75, 210
Consistency 5, 12, 15, 40, 43, 116
    in color–music synesthesia, 105
    as fallible test, 83–84, 237
    in personification synesthesia, 83–84
    as test of synesthesia, 51, 64–65, 218,
        224, 235
Contextual influence, 13, 68, 70, 74–
    75, 75f, 106–107, 116, 135, 211,
    231–232, 241–242
Contour, 3f, 32, 165
Continuum, cognitive, 13, 166, 196
Contrast dependence, 77–78, 210
Conveyor belt analogy, 203
Cool glass columns, 18–19, 59
Correspondences. *See* Cross-sensory
    mapping
Correspondences (poem), 188
Crane, Carol, 45–46, 173
Creativity, 7, 102, 163–198, 226, 237
Crick, Francis, 235
Critical periods, 35, 214, 228
Critical phonemes (in taste), 37, 147–
    148, 147f
Cross-sensory mapping, 43, 103–105,
    134–136, 196. *See also* Ventriloquist
    illusion
    expansion of maps, 204–205
    in infants and children, 106–107

nonmodal, 106
orderliness of, 90, 104, 104f,
163–164
Cross talk, 6, 164, 197
in normal brain, 105–107, 214–216
in synesthesia, 199, 205–209, 232
Cytoarchitecture, 230
Cytowic, Richard, 1–3, 7, 9, 20,
47, 135, 138, 156, 183, 224,
244–245
Cyto sculpture, 178–179, 179f
shape of name, 88

Dance, 164
Dann, Kevin, 188
Days of week, 8, 38f, 110
Day, Sean, 24f, 43, 87, 171–172, 235
as synesthete, 23, 55, 77, 82, 140
De Caprio, Tony, 93
De Maistre, Roy, 188
Deactivations, 136
Dehaene, Stanislas, 123
Deliberate contrivances, 13, 187–194
Delville, Jean, 192f
Depth electrodes, 217
Dermatome chart, 133, 133f
Development, 13–14, 38t, 40, 90,
99–101, 105–106
differentiation vs. integration, 107
language vs. experience, 142,
172, 197
Dexter, John, 185–187
Diagnostic features, 19, 47–56
Diet in tasted words, 148
DiRaimondo, Linda, 81
Direct realism, 246–247
Directionality, 103, 127–129, 146,
240–241
Disney, Walt, 90, 191
Dolby, Kristen, 145, 150
Dolphin visual-token sign
language, 197
Doty, Richard, 136

Downey, 140–141
Duffy, Pat, 111f
Dynamic quality, 28, 44
Dyscalculia, 243

Eagleman, David, 9, 25, 225
Early stages of perception, 48–49, 106
Ebbinghaus illusion, 241–242, 242f
Eidetic memory, 53–54, 130, 146
in Nabokov, 54
Elemental nature of percept, 51–52
Emotion, misattributed, 154, 160
Emotional coloring, 45, 54–56, 151–
154, 158
affecting wardrobe choices, 55
Emotional intelligence, 152
Emotionally mediated synesthesia,
35–36, 151–157, 250
Emrich, Hinderk, 8
Entopic phenomena, 57
Epigenetic, 23
Epileptic synesthesia, 222–223
Epiphenomenon, 160
Erogenous sensations, 130, 133, 159

Fantasia, 90, 191–193
Fantasy proneness, 194
Fechner, Gustav, 17
Feedback connections, 202–204, 228
Feeling of a presence, 160
Feelings vs. emotion, 158
Female bias. *See* Sex ratio
Fetal brain, 214
Feynman, Richard, 28
Fischinger, Oskar, 190
Finger agnosia, 243–244
Fingertip agraphesthesia, 244
Firework analogy, 14, 39, 81, 87, 143
First-letter effect, 32–33, 37, 40, 67–68,
79, 146, 148
First-letter priming, 79
First-person vs. third-person accounts,
16, 19, 20

Flavor (as opposed to taste), 127, 131
changed by music, 142–143
Folk psychology, 17, 156
Forced thinking, 160–161
Foreign alphabets, 67
Form constants, 18, 51, 56–61, 58f–60f,
136, 174f, 178, 184, 218
Fusiform gyrus (left), 150, 210

Galton, Sir Francis, 5, 7, 10, 26–28,
26f, 27f, 51, 117, 224
Galvanic skin response, 85
Gene expression, 145, 197, 226
Genetics. See Synesthesia
Geometric percepts, 14, 43–44, 57,
157, 178, 186
Geschwind, Norman, 196–197
Golden Gate bridge, 151–152
Goosebumps, 45
Graphemes. See also Colored
graphemes
age of recognition, 36
grapheme–phoneme conversion, 36
graphemes vs. phonemes, 32
Glaucoma, rainbows in, 59
Global neural network, 156, 190,
202–203, 213, 232
Grossenbacher, Peter, 215–216
Gut feelings, 158

Haiku, 189–190
Hallucinations, 17–18, 56–57, 71–73,
217–219, 223, 239
Hashish, 188
Hebbian rule of learning, 216
Hidden patterns, 48–49, 74,
Higher vs. lower synesthesia, 77
Hirshhorn, 193
Historical number of publications, 6,
16f, 25f
Hockney, David, 13, 173, 183–186
Homeostasis, 158
Homographs, 67–69, 69t, 101

Homonyms, 32, 33t, 148
Homunculus, 131–132, 132f
Hubbard, Edward, 8, 48, 75, 77,
124, 150, 156, 163–165, 195, 197,
207–208, 210–211, 213, 226
Huysmans, Joris-Karl, 142

Iconic thinking, 172
Imagery, failure of, 206–207
Implicit learning, 129
Imprinting, 79, 117, 229
Ineffability, 18, 56
Influences on cognition, 243–246
Infrared methods, 20, 50, 205
Internalized speech, 172
Internested forms, 112–113
Interpretation of perceptions, 17–18,
56–57
Introspection, 15

James, William, 172

Kaleidoscopic effect, 39, 61, 87, 219
Kandinsky, Wassily, 13, 59, 98–99,
101, 176, 187
Karamzin, Nikolai, 250
Key (musical), 56, 90–92, 94, 97, 103,
190, 191f, 229
Kindling, 161
Kissing as synesthetic trigger, 159
Klangfarbe, 91
Klee, Paul, 188
Klüver, Heinrich, 18, 56–58, 178, 218

Lakoff, George, 166
Laszlo, Alexander, 190
Learned associations, failure of, 206
Learning, role of, 30, 35, 37, 79–80,
118, 228
L'Enfant et les Sortilèges, 183–184
Les sonnet des voyelles, 188–189
Levels of processing, 73–77
Levitin, Daniel, 251

Lexemes, 37, 146
Ligeti, György, 93
Limbic vs. nonlimbic associations, 196–197
Lipreading, 106
Liszt, Franz, 93, 173
Localization, failure of, 199, 208, 232
Localizers vs. nonlocalizers, 71–73, 77, 87–89, 109, 212
Lolita, 253
Long, Joseph, 96–98, 229
Loomis, Marnie, 154
Lorch, Tammy, 114
Lovelace, Chris, 216
Lower vs. higher synesthesia, 77
LSD, 4, 47, 140, 216–218
Luria, A. R., 4, 53, 80, 142–143, 146, 246

Magnitude comparisons, 28, 138, 240–241
Manohar, Sanjay, 77
Man Who Tasted Shapes, 3, 221
Mapping. See Cross-sensory mapping
Marks, Larry, 103–104, 163–164, 164f, 171
Maurer, Daphne, 214
McGurk effect, 106, 203, 242
Meditative states, 172, 217, 220–221
Meehan, Susan, 157, 159
Melody, 91, 184
Memorability, 52–53
Memory aid, 80–81
Memory explanation of synesthesia, 5, 79
Merikle, Phil, 211
Mescaline, 418, 56
Messiaen, Olivier, 13, 93–95, 105, 173–175, 184, 187
  composition method, 94
  musicological analysis of, 94–95
  stained-glass effect, 94, 175
Metamorphopsia, 218

Metaphor, 7, 163–165, 194–196
  conflicts among, 169
  development of, 169–172, 195–197
  ontological, 167–169, 171, 196
  orientational, 166–167, 196
  physical basis of, 166, 169
  polarity of, 171, 194
Metonymies, 87, 91–92, 171
Milogav, Jean, 2, 50, 55–56, 74
Mind of a Mnemonist, 4, 53, 246
Misdiagnosis, 10
Misattribution of affect, 160
Modularity, 105, 150, 231–232
Morrow, Mike, 88
Mousatkis, Mathew, 37
Movement (as synesthetic trigger), 46–47
Movement (of synesthetic percept), 14, 28
Muscae volitantes, 57, 59
Music in restaurants, 142
Musical harmony, 99
Musical instruments, 45
Musical interval, 91, 104–105, 143, 144t
Musical notation, 32, 190, 192f

Nabokov, Dmitri, 11–12, 157, 175, 249–254
Nabokov, Vera, 11
Nabokov, Vladimir, 13, 173, 175f, 176, 176f
  poem on synesthesia, 176
  as synesthete, 2, 10–12, 36, 40, 54, 227
Nature vs. nurture, 37, 228–229
Navon figures, 75, 76f, 84, 84f
Negative numbers, 119–120
Neural circuit for synesthesia, 208
Neural localization of function, 25f, 199–202

Neural networks for synesthesia,
  99–203, 209, 212
  network of networks, 202
Neural outgrowth, 214
Neural overgrowth, 204–205
Neural pruning, 204–205, 214
Nicotine, 138
Nolan, Murial, 140, 245
Nonmodal senses, 107
Nonwords, 37, 38f, 146, 148–149
Normality of synesthesia. See
  Synesthesia
Number forms, 2, 26–30, 90, 109,
  140. See also Spatial sequence
  synesthesia
  change with age, 114, 116, 120
  common motifs, 117–122
Number lines, 106, 109, 115–117,
  123–124, 212, 238
Nunn, Julia, 73, 206–207, 206f

Oedipus Rex, 184–186
Odor priming, 130
O'Keeffe, Georgia, 13, 187–188
Olfaction, 96, 127–128
  selective loss of, 96
Omission of certain colors, 66,
  238–239
On the Spiritual in Art, 101
Operculum, 128
Orange Sherbet Kisses, 59, 102, 159, 251
Orgasms, 157–159
Out-of-body experience, 160

Pain (as synesthetic trigger), 43–44, 180
Pale Fire, 54
Palinopsia, 218
Panoramic viewing, 28, 30, 109
Parade, 185
Paradigm shift, 21, 230
Paradox, tolerance for, 169
Parallel processing, 201
Parietal lobule, inferior, 196

Peak popularity, 15
Perfect pitch, 91, 95–98
  similarities to synesthesia, 96
Penetrance, 225–226
Peripersonal space, 50
Perlman, Itzhak, 93
Personification, 26, 40–43, 82–85, 227
Phonemes, 5, 36, 87, 136–137
  age of recognition, 35–36
  colored, 99–101
  tasted, 5, 37, 77, 101–103, 145–150
Photisms, 39, 72, 151
  depiction of, 177–179
  size determinants, 93
  sound-induced, 219, 219t
Photographic memory. See Eidetic
  memory
Pike, Marti, 28–35, 40, 110–120
Pitch (musical), 87, 91–92, 96–98
Plasticity, 98, 214, 216, 228
Polyphony, 188
Pop-out, 49
Portentousness, feeling of, 152, 160
Prajnaparimita, 221
Prevalence, 7–8, 25, 47, 54, 122–123
Price, Rebecca, 39, 87, 141–142
Priming. See Color priming; First-
  letter priming; Odor priming
Projection, 159–161
Prometheus, 190, 192f
Proverb interpretation, 195
Proximity theory, 119, 150,
  212, 239
Pruning, 204–205, 214, 232
Pseudo-synesthesia. See Deliberate
  contrivances
Psychic seizures, 161, 222
Psychophysics, 17
Publications, historical number
  of, 6, 16f, 25f
Pupil diameter, 50
Pushkin, Aleksandr, 250
Pythagoras, 82–83

Qualia, 14, 87, 196

Rain Man, 241
Rake's Progress, 184
Ramachandran, Vilayanur, 46, 75, 82–83, 93, 102, 122, 127, 156, 188–189, 197–198
Random dot stereograms, 54
Raven's progressive matrices, 198
Recipes, synesthetic, 131
Refrigerator magnets, 5, 79, 216, 229
Release hallucinations, 218–220
Rhinencephalon, 127
Right–left confusion, 224
Rimbaud, Arthur, 176, 188–189
Rimsky-Korsakov, Nikolai, 93
Rossignol, 183, 186–187
Roxburgh, Julie, 102

"S" (Luria's nmemonist), 51–52, 122, 142, 145–148, 246
Saenz, Melissa, 46
Sagiv, Noam, 82–84, 122
Sanders, Dana, 197
Sandokai, 221
Savants, 221
"Scent of color," 98
Scientific American, 190
Schoenberg, Arnold, 101, 176
Schultze, E., 142–143
Screen, projection onto, 13–14, 39, 50, 88, 177
Scriabin, Alexander, 13, 187, 190–192
Scripture of great wisdom, 221
Seizure. See Temporal lobe epilepsy
Selection pressures, 239
Semantic meaning, 39, 47, 52, 81, 145, 148
Sense of direction, 243–246
Sensory confusion, 14, 68, 70f
Sensory coupling, 110–111, 197
Sensory deprivation, 39, 57, 217–220

Sensory overload, 55, 102–103, 143, 146, 217, 243
Sensory substitution, 205
Sequence (ordinality), 109, 224–226
Sex ratio, 9, 23, 96, 103
Shanon, Benny, 79
Sibelius, Jean, 93
Silva, Colleen, 111t, 114f, 120, 121f, 238
Simner, Julia, 8–9, 23, 52, 78–79, 122, 145–147, 224–226
Simon, Deni, 10, 14, 51, 154, 157, 159, 243
Simpson, Lidell, 46, 103, 17
Shades, multiplicity of, 51–52
Slaughterhouse Five, 124
Smell, 127–129, 129f. See also Olfaction
Smell Identification Test, 136
Smilack, Marcia, 46, 102, 110, 112, 115, 119, 125, 180–182
Smilek, Dan, 80–81
Smith, Laurel, 88, 92, 98, 100–101, 177–181
Smith, Sherelle, 56
Sonar input, 205
Son et lumière, 13, 187
Sound triggers, 14, 39, 85, 143
Spatial extension of perception, 50, 56, 59, 72–73
Spatial memory, 76–77, 245
Spatial quality of color, 185
Spatial representation, 119, 212–213
Speak, Memory, 2, 54
Specificity of descriptions, 37, 51–52
SNARC effect, 123–125
Steen, Carol, 1, 59, 74, 177, 179f–181f
Steller's jay, 94
Stream of consciousness, 172, 209
Stroop interference, 47–48, 76, 84, 104, 144, 149, 154
   cross-sensory version, 48, 104–105
   gustatory version, 144
Subjectivity, 15, 17, 236

Synesthesia, 4–5
  age of onset, 9–11, 38t, 152, 214,
    221, 229
  bidirectional, 94–95, 102–103, 142,
    194, 243
  change/loss with age, 9–12, 64–65,
    149, 251
  childhood, 1, 12, 63–64, 172, 214
  cognitive deficits in, 243–246
  in concussion, 222
  consistency (see Consistency)
  definition, 1, 109–110, 112
  diagnostic features, 47ff
  directionality of, 127–129, 146,
    240–241
  drift over time, 67–71, 68f, 152,
    237–238
  drug effects on, 137–140, 139f,
    224
  drug-induced, 5, 56, 172
  familial nature, 5, 8–11, 224
  fivefold, 4, 53, 80, 143
  frequency of types, 23–25, 24t, 43,
    171–172
  genetics, 9, 96, 224–229, 237
  heterogeneity of, 227
  hidden importance, 194–195
  historical publications, 6, 16f, 25f
  idiosyncrasy of, 4–5, 17, 176
  in literature, 2, 54, 253
  lower vs. higher, 39, 77, 87, 140–
    142
  nonsensory couplings, 25
  as normal phenomenon, 13, 66,
    103–108, 203–204, 216
  occupations, 173
  polymodal, 23, 37–39, 45–46, 77, 90,
    98–99, 109, 122
  qualities of, 40, 43, 47, 53, 64,
    145–150
  simple (see Emotionally mediated)
  skeptical criticism of, 4–5, 205
  spatial sequence, 25–30, 109–125

  stability, 40–43
  tests for, 25 (see also Consistency)
Synesthesia Battery, 25
synesthesia euphoria, 190
Synesthesia List, 23, 173
Synesthetic art. See Art

Tapestry analogy, 34
Taste, 5, 37, 93, 127–150, 143–149.
    See also Flavor
  influenced by color, 163–164
  influenced by smell, 127–129
Tasted words, 5, 37, 77, 101–103,
    145–150
Temporal lobe epilepsy, 222–224
  taste and smell sensations, 222–224
Test–retest. See Consistency
Texture of reality, 20–21
Theories of synesthesia, 214–216
  decreased inhibition, 198, 215–216
  increased wiring, 214–215
Theory of mind, 152
Theosophy, 190
Ticker tape effect, 50, 74
Timberlake, Megan, 40, 41t, 74
Timbre (music), 45, 87, 91–93, 98–99,
    105f, 142, 144
Time sequences, judgment of, 47
Tongue, as brain–machine interface,
    205
Tonotopic cortex, 90–91, 211
Torke, Michael, 93
Trails, 218
Transparent overlay, 213
2001: A Space Odyssey, 93
Types of synesthesia
  audio-motor, 40, 43, 103
  colored days of the week, 8
  colored graphemes, 7–8, 12, 37
  colored hearing, 14, 23–25, 63–85,
    88–95, 142–145, 151–161,
    214–215
  colored phonemes, 51

colored smell, 130–136, 140–142
color–music, 11–12
emotionally mediated, viii, 35–36,
   151–152
flavor–touch, 23
geometric pain, 45
nonrandom, 171
number–form, 19, 26–30
personification, 55, 98–99
smell–shape, 239–240
sound–movement, 146
sound–personality, 240
sound–touch, 40–47
sound–visual, 227
spatial sequence, 46–47, 237–238
tasted color, 43, 81–85
tasted voices, 145
tasted words, 5, 37, 77, 101–103,
   145–150
Tyutchev, Fyodor, 253

Ugly colors, 6
Universal correspondences, search
   for, 101, 176
Unpleasant synesthesia, 102, 143.
   See also Sensory overload
Unusual experiences, 159f
Upside-down clock, 32f

V4, 6, 30–36, 45, 73, 204–212, 239
Valence, 156
van Halen, Eddie, 93
Ventriloquist illusion, 37, 106, 164–
   165, 203
Vibraphone, 51
Virtual reality, 30, 115–116
Vision, multiple aspects of, 240
Visitor experience, 161
Visual field, periphery, 161
Visual qualia, 87
Visual word form area (VWFA),
   210–212
Voice, 39, 181

Volume transmission, 231
Vonnegut, Kurt, 124

Wagner, 98
Walsh, Roger, 220–222
Wannerton, James, 5, 37, 146–149
Ward, Jamie, 104, 145, 147–148, 154,
   156
Warm vs. cool colors, 169–170
Watson, Michael, 3, 18–20, 23, 50,
   57–59, 93, 130, 133–134, 136, 139,
   159, 217, 224, 243
Wonder, Stevie, 93
Word frequency, 148–149, 154
Wood, Thomas, 93

X chromosome, 9, 224–225
Xenon regional cerebral blood flow,
   136

Zimmer, Bill, 59
Zooming in, 30, 109, 214